Lecture Notes in Mathematics

Volume 2287

This series reports on new developments in all areas of mathematics and their applications - quickly, informally and at a high level. Mathematical texts analysing new developments in modelling and numerical simulation are welcome. The type of material considered for publication includes:

1. Research monographs
2. Lectures on a new field or presentations of a new angle in a classical field
3. Summer schools and intensive courses on topics of current research.

Texts which are out of print but still in demand may also be considered if they fall within these categories. The timeliness of a manuscript is sometimes more important than its form, which may be preliminary or tentative.

Titles from this series are indexed by Scopus, Web of Science, Mathematical Reviews, and zbMATH.

More information about this series at http://www.springer.com/series/304

Björn Gustafsson • Yu-Lin Lin

Laplacian Growth on Branched Riemann Surfaces

 Springer

Björn Gustafsson
Department of Mathematics
KTH Royal Institute of Technology
Stockholm, Sweden

Yu-Lin Lin
Deparment of Mathematics
University College London
London, UK

ISSN 0075-8434   ISSN 1617-9692   (electronic)
Lecture Notes in Mathematics
ISBN 978-3-030-69862-1   ISBN 978-3-030-69863-8   (eBook)
https://doi.org/10.1007/978-3-030-69863-8

Mathematics Subject Classification: Primary: 30C20; Secondary: 76D27

This Springer imprint is published by the registered company Springer Nature Switzerland AG.
The registered company address is: Gewerbestrasse 11, 6330 Cham, Switzerland

Picture from the Banff International Research Station, programme "Integrable and stochastic Laplacian growth in modern mathematical physics" (organized by D. Crowdy, B. Gustafsson, J. Harnard, M. Mineev, and M. Putinar), October 31–November 5, 2010. From left: Michiaki Onodera, Makoto Sakai, Björn Gustafsson, and Yu-Lin Lin

In memory and honor of Makoto Sakai (1943–2013).

Preface

This book grew out of investigations on the Polubarinova-Galin equation, describing the conformal map onto a growing Hele-Shaw fluid blob, by one of the authors (Lin); see [69–72]. This was followed up by joint work by Gustafsson and Lin [35] concerning the dynamics of zeros and poles of the derivative of the conformal map. The Hele-Shaw problem also has a description in terms of weak solutions related to balayage, more precisely partial balayage, this being much related to quadrature domains for subharmonic functions. These tools have been developed by Makoto Sakai [97, 98] partly in collaboration with one of the authors [41]. More recently, the tool of partial balayage was, in collaboration with Joakim Roos, extended to Riemannian manifolds [40].

The present book can be viewed as a continuation and extension of the work by Gustafsson and Lin [35] to the case of conformal maps that are no longer univalent in the usual sense but then still can be viewed as conformal maps into Riemann surfaces that are branched coverings of surfaces over the complex plane. For this purpose, the tools of Gustafsson and Roos [40] are needed. The original aim was to prove the existence of a global in time solution starting from an arbitrary conformal image of the unit disk. On the one hand, this turned out to be more difficult than expected (there still remains a technical obstacle to be overcome), but on the other hand the subject as a whole has showed up to be extraordinarily rich and interesting from different points of view. We have in this respect taken the opportunity to connect the Hele-Shaw problem or, by another name, Laplacian growth, to some recent developments in mathematical physics, relating to integrable hierarchies, string equations, and Hamiltonian systems in infinitely many time variables. This material is inspired by the papers of Igor Krichever, Mark Mineev-Weinstein, Paul Wiegmann, Anton Zabrodin, and others.

Many topics in the book are close in spirit to the work of Makoto Sakai. One of the authors of this volume collaborated with Sakai for more than twenty years, but very sadly Sakai passed away, unexpectedly, in December 2013. Sakai started working on extremal problems for analytic functions on Riemann surfaces in the late 1960s. For example, in his first published paper [93] Sakai solved an open problem stated in the work by Sario and Oikawa [106]. Continued work in this direction led him to tools of potential theory for constructing extremal functions and corresponding domains. The appendix in the mentioned Lecture Notes [97] gives an excellent account for how things are connected. See also [95, 96], for example. Starting from a different extremal problem for analytic functions, Dov Aharonov and Harold S. Shapiro [2] were led to similar kinds of extremal domains, and they named them quadrature domains. This terminology is now generally adopted. See [42, 109] for overviews.

These quadrature domains are, via a discovery by Stanley Richardson [88], linked to the Hele-Shaw problem, as explained in the work by Sakai [97]. See also [98] where quadrature domains are constructed in essentially the same way (namely, by using variational inequalities), as weak solutions of the Hele-Shaw problem are constructed in the work by Gustafsson [29]. A related construction appears in the work by Elliott and Janovsky [18].

Almost all papers (40 published papers in total, according to MathSciNet) by Makoto Sakai concern directly or indirectly topics connected to quadrature domains. In one of his least known papers [100], Sakai discusses quadrature Riemann surfaces. That paper is particularly close to the present Lecture Notes. In that paper, Sakai got stuck on one technical point, which he left open, and that point is essentially the same as our technical difficulty, briefly mentioned above.

We would like to dedicate this book to the memory of *Makoto Sakai* (1943–2013). Sakai made original and groundbreaking contributions in potential theory, highly relevant for the material in the present treatise. He was one of the creators of the theory of quadrature domains, and his books [97, 105] on the subject will have a long-standing impact on the development of the subject, and on potential theory and its applications in general. His papers and books are not always easy to read, but they are very sharp, and Sakai's work is now receiving increasing recognition in the general mathematical community.

Stockholm, Sweden   Björn Gustafsson
London, UK   Yu-Lin Lin
January 18, 2021

Contents

Chapter 1
Introduction

Abstract This chapter gives an overview of the main themes of the book.

1.1 General Background

Laplacian growth refers to evolution of domains, in the complex plane for example, for which the boundary propagates in the normal direction with speed proportional to the normal derivative of the Dirichlet Green function with pole at a given source (or sink) point. Equivalently, the speed is proportional to the harmonic measure of the boundary. The domains may be expanding or shrinking, depending on the sign of the constant of proportionality. The above description of Laplacian growth easily extends to more general settings, for example to higher dimensions, to Riemannian manifolds, to more general source (or sink) configurations, to cases with the Laplace equation being replaced by other elliptic equations, *et cetera*. In this introductory chapter we give a brief survey of the subject, and a rather extensive overview of those topics that will be covered by this book. A few general books and survey papers on the subject, and some of its ramifications, are [43, 47, 59, 78, 119, 126].

There is a huge difference between the case of expanding domains and that of shrinking domains in Laplacian growth. The case of expanding domains is highly well-posed in the sense of stability of solutions, and also on that the regularity of the boundary in that case improves as time goes on. The case of shrinking domains is, on the other hand, severely ill-posed, and not even local in time solutions exist unless the initial boundary is analytic. This ill-posed version is actually the most studied case, often in the reversed geometry with unbounded domains and a sink at infinity, and sometimes with stochastic ingredients. The present book mainly focuses on the well-posed case (expanding domains), but partly with an eye on getting insights into the ill-posed case. See brief discussions in Sects. 7.1.4 and 7.2.

Laplacian growth models a large number of processes in nature. The original source for most of the early papers on the subject is the evolution of a blob of viscous fluid in the narrow gap between between two parallel planes. The history of this problem started with an experiment and a subsequent paper [54] by Henry Selby Hele-Shaw in 1898, and Laplacian growth therefore often bears the name

© The Author(s), under exclusive license to Springer Nature Switzerland AG 2021
B. Gustafsson, Y.-L. Lin, *Laplacian Growth on Branched Riemann Surfaces*,
Lecture Notes in Mathematics 2287, https://doi.org/10.1007/978-3-030-69863-8_1

Hele-Shaw flow, more exactly with a free, or moving, boundary. In this volume Hele-Shaw flow (with a source or sink term) and Laplacian growth will be treated as synonymous concepts. The mathematical equations describing the flow, including a reduction of dimension from the original three physical dimensions to the two dimensional governing law for the fluid blob, were derived by H. Lamb in an early edition of [67].

In the 1940s the subject was taken up by scientists in the Soviet Union, in particular L. A. Galin, P. Ya. Polubarinova (Polubarinova-Kochina after marriage with N. Ye. Kochin), Yu. P. Vinogradov, P. P. Kufarev. Important publications from this period are [24, 85, 121]. See in general [120] for the interesting history of the subject and of the scientists who have developed it. Starting in the 1950s Great Britain became a center of development, with work by P. G. Saffman, G. I. Taylor, S. Richarson, J. Ockendon, S. Howison and many others. The papers [88, 92], in the period up to 1980, have been particularly important for the subject. For the development after 1980 we refer to [120] (see also [9]) and the previously mentioned books and surveys.

The basic setting in the present book is Laplacian growth of simply connected domains in the complex plane, and extensions of this to branched Riemann surfaces. When the fluid domain, generically called $\Omega(t)$, is simply connected it is natural to describe it in terms of the (time dependent) conformal map $f(\cdot, t)$ from the unit disk $\mathbb{D}$, subject to the standard normalizations $f(0, t) = 0$ and $f'(0, t) > 0$. Implicit here is that the source point has been chosen to be the origin in the complex plane. The equation to be satisfied by $f(\zeta, t)$ was found by Polubarinova and Galin (independently) and bears their names. It reads

$$\text{Re}\left[\dot{f}(\zeta, t)\overline{\zeta f'(\zeta, t)}\right] = 1 \quad \text{for } \zeta \in \partial\mathbb{D}, \tag{1.1}$$

and it expresses exactly that the speed of the boundary $\partial\Omega(t)$ in the outward normal direction equals 2π times the harmonic measure, or normal derivative of Green function, with respect to the origin. This is seen by interpreting the left member as $|f'|$ times the inner product between $\dot{f}$ and the outward unit normal vector $\zeta f'/|f'|$.

Equation (1.1) can be solved for the time derivative $\dot{f}(\zeta, t)$, and the result is what we call the Löwner-Kufarev equation, which gives $\dot{f}(\zeta, t)$ essentially as a Poisson integral:

$$\dot{f}(\zeta, t) = \frac{\zeta f'(\zeta, t)}{2\pi i} \int_{\partial\mathbb{D}} \frac{1}{|f'(z, t)|^2} \frac{z + \zeta}{z - \zeta} \frac{dz}{z} \quad (\zeta \in \mathbb{D}). \tag{1.2}$$

If $\partial\Omega(0)$ is analytic in the sense that $f(\cdot, 0)$ is univalent in a full neighborhood of the closed unit disk, then (1.1) and (1.2) have a unique solutions in some interval $-\varepsilon < t < \varepsilon$ (see [19, 71, 87, 115, 121] for proofs). If $\partial\Omega(0)$ in addition is star-shaped with respect to the origin, then it can be shown (see [36]) that the solution never breaks down, hence exists for all $-\varepsilon < t < \infty$. However for more general initial shapes different parts of $\partial\Omega(t)$ may go into collision and univalence of $f(\cdot, t)$

may be lost. The problems which then appear represent the main themes to be discussed in the book.

1.2 Loss of Univalence, Several Scenarios

There are several ways to treat loss of univalence. These will involve both the Löwner-Kufarev and the Polubarinova-Galin equation for the conformal map, as well as weak solutions in the form of families of domains, which may become multiply connected. The present section with its illustrations (Figs. 1.1 and 1.2) can be viewed as a general synopsis of the central themes in the book.

We start with a conformal map $f(\cdot, 0)$ from the unit disk onto a domain $\Omega(0) \subset \mathbb{C}$ which is close to becoming multiply connected. See Fig. 1.1, bottom view. For definiteness we may think of f being a polynomial, and it is actually enough to take a polynomial of degree three in order to obtain the evolutions to be described. Thus, as a leading example, we start at time $t = 0$ with

$$f(\zeta) = a_0\zeta + a_1\zeta^2 + a_2\zeta^3, \tag{1.3}$$

and then this polynomial form of f will be stable in time (with $a_j = a_j(t)$), at least for a short period. With $\Omega(0)$ looking as in Fig. 1.1 there will come a time $t = t_1 > 0$ when two parts of $\partial\Omega(t)$ touch each other. At that time one has to choose between two ways to continue.

(1) One may allow $\Omega(t)$ to become multiply connected. This choice is represented by the pictures to the right. Then one looses contact with the conformal map and the Eq. (1.1). But it turns out that there is a good concept of weak solution (sometimes called variational inequality weak solution), based on integrating (1.1) with respect to time. This leads to the property, holding for solutions of (1.1), that for any $t > 0$ for which the solution exists, and for any function h which is harmonic and integrable in $\Omega(t)$,

$$\int_{\Omega(t)} h\,dxdy - \int_{\Omega(0)} h\,dxdy = 2\pi t\,h(0). \tag{1.4}$$

A slightly stronger property actually holds, that allows subharmonic test functions above provided the equality sign $=$ is replaced by the inequality $\geq$. This expresses then that $\Omega(t)$ is a quadrature domain [42] for subharmonic test functions and with respect to the measure $2\pi t\delta_0 + \chi_{\Omega(0)}$.

In the example with $f(\cdot, 0)$ on the form (1.3), $\Omega(0)$ is itself a quadrature domain, admitting a quadrature identity of the form

$$\int_{\Omega(0)} h\,dxdy = c_0h(0) + c_1h'(0) + c_2h''(0) \tag{1.5}$$

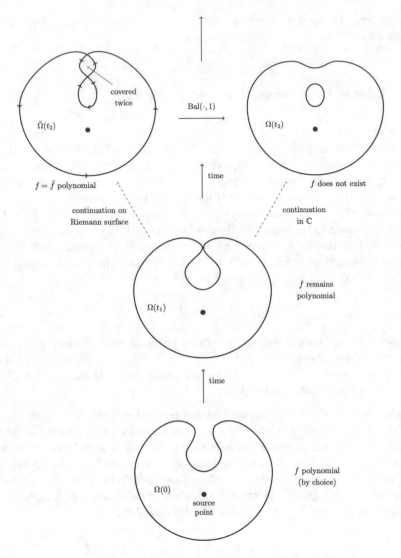

Fig. 1.1 Two kinds of evolutions starting from a simply connected domain ($0 < t_1 < t_2$)

for analytic test functions h (we let h be analytic here just to give the derivatives a definite meaning). Inserting this into (1.4) it follows that $\Omega(t)$ satisfies a similar quadrature identity, with c_0 depending on t.

The statement of $\Omega(t)$ being a quadrature domain for subharmonic functions can be reformulated as saying that $\chi_{\Omega(t)}$ is the result of performing a certain

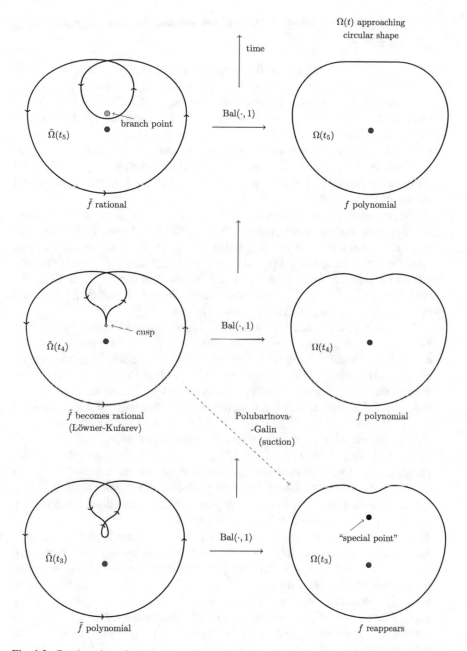

Fig. 1.2 Continuation of previous pictures $(0 < t_1 < t_2 < t_3 < t_4 < t_5)$, including a possible suction phase

partial balayage ("partial sweeping") of the measure $2\pi t \delta_0 + \chi_{\Omega(0)}$ to density one. We write this as

$$\chi_{\Omega(t)} = \mathrm{Bal}\,(2\pi t \delta_0 + \chi_{\Omega(0)}, 1). \tag{1.6}$$

This partial balayage can be viewed as an orthogonal projection in a Hilbert space, and in particular it follows that $\Omega(t)$ can be constructed directly out of $\Omega(0)$, without using any intermediate domains. Equation (1.6) defines then a global ($0 \le t < \infty$) weak solution in form of a family of domains $\Omega(t)$, unique up to nullsets but possibly multiply connected during certain stages. This is the evolution in the pictures to the right in Figs. 1.1, 1.2. Weak solutions in general are discussed in Chap. 3.

(2) The other way to continue from $\Omega(t_1)$ is to observe that an initial non-univalence of f may remain unnoticed by the Eqs. (1.1) and (1.2). Not until zeros of f' reach $\partial \mathbb{D}$ will these equations be affected. So the solution can go on as "wrapped Hele-Shaw flow" on a Riemann surface for a while. However, it turns out that whenever f has become non-univalent then one is in a trapped situation: zeros of f' will inevitably reach $\partial \mathbb{D}$, sooner or later (see [35]). And then the Eqs. (1.1) and (1.2) are in serious trouble.

The two ways, (1) and (2), of treating loss of univalence are illustrated in Fig. 1.1 at the time level $t = t_2$. The domain to the left is wrapped and has multiplicities, while the domain to the right has a hole. These pictures in general are only schematic, and are moreover scaled so that they all have the same general size. In reality the domains actually swell with increasing time, and for the correct sizes one also has to take into account that parts of the domains to the left are covered twice. The arrows on the boundary curves to the left should indicate the mapping degrees (or index) for the various pieces.

The two evolutions, (1) (to the right) and (2) (left), are connected via partial balayage: the single-sheeted domain to the right can be obtained from the domain to the left by "smashing" its multiplicities to density one by $\mathrm{Bal}(\cdot, 1)$, as indicated. Thus the arrows in the upper part of Fig. 1.1 constitute a commuting diagram. This is discussed and proved in Sect. 4.4.

The next time, $t = t_3$, illustrated in Fig. 1.2 represents a phase change in the evolution of the weak solution to the right: the hole in the doubly connected domain has shrunk to a point, and adopting this point (called a "special point" in the context of quadrature domains, see [42]) to the domain one has again a simply connected domain. The quadrature identity (1.5) remains valid for $\Omega(t)$, with $c_0 = c_0(t)$, and therefore the conformal map is again a third degree polynomial. The special point, as well as the hole in the previous stage, is called a "droplet" in related contexts of weighted equilibrium configurations, random matrix theory, Coulomb gas ensembles (etc). See for example [50, 65, 127] for these matters.

In the pictures to the right the evolution then goes on smoothly for all $t_3 < t < \infty$, in the present example. The evolution is real analytic with respect to time except at moments of phase change, and the boundary curve is algebraic at each time (this is a general property for quadrature domains of the form (1.5)). In the present treatise we express this algebraicity via the exponential transform and elimination function, see Sects. 3.5 and 3.6. Asymptotically $\Omega(t)$ approaches circular shape as $t \to \infty$, see for example [70, 81, 82, 104], or else use the "inner normal theorem" [43].

The evolution in the left pictures continues to develop smoothly as wrapped Hele-Shaw flow parametrized by the conformal map until, at a certain time $t = t_4$, a cusp develops on the boundary. This means that a zero of f' has reached $\partial \mathbb{D}$, and this is the most critical moment in the evolution. Somewhat remarkably, it is possible to let the zero of f' pass through $\partial \mathbb{D}$. Then f will map $\mathbb{D}$ into a branched Riemann surface, and we sometimes rename f to $\tilde{f}$ when it is viewed that way. In terms a uniformizing coordinate on the Riemann surface, resolving the branch point, the evolution becomes weighted Laplacian growth. Compare Figs. 4.1, 4.2, 4.3, which represent a somewhat simpler case. The passage of zeros of f' becomes more smooth if one relaxes (1.1) by replacing the right member by a time dependent source strength $q(t) \geq 0$, which at critical moments is allowed to vanish. Similarly for (1.2).

With some zeros of f' inside $\mathbb{D}$, but none on $\partial \mathbb{D}$, the Eqs. (1.1) and (1.2) still make sense. However, they are then no longer equivalent. Equation (1.2) is the stronger one, having a unique solution, while (1.1) will admit finite dimensional families of solutions. In general these can be parametrized by the branch points

$$B_j = f(\omega_j, t),$$

corresponding to the zeros $\omega_j = \omega_j(t)$ of $f'(\cdot, t)$ inside $\mathbb{D}$. Thus, for solutions of (1.1) the branch points depend on t and can move freely. The unique solution of the Löwner-Kufarev equation (1.2) is then characterized by the B_j being stationary. This means that f can be viewed as a map into a fixed branched Riemann surface. Detailed discussions are found in Chap. 2, and also on in Chaps. 8, 9, 10.

All this is illustrated in the top left picture with $\tilde{\Omega}(t_5)$. The zero of f' has been swallowed by $\mathbb{D}$ and f maps $\mathbb{D}$ onto a Riemann surface with one branch point. The evolution represents a solution of the Löwner-Kufarev equation, this solution being unique and hence "canonical". Since the branch point does not move it is the same point as the cusp point at time $t = t_4$. After time $t = t_5$ no more drastic things need to happen.

(3) Besides the above two main tracks, there is also a side track that we want to mention. This is represented by the dashed arrow in Fig. 1.2 and starts from the domain $\tilde{\Omega}(t_4)$ in Fig. 1.2, which has a cusp point and a piece which is covered twice. Starting with f of the form (1.3) it turns out that the Polubarinova-Galin equation admits a global ($0 < t < \infty$) solution on the same polynomial form, and this solution is real analytic in t provided the source strength $q(t)$ is suitably adapted. In particular $q(t)$ need to vanish for $t = t_4$, and after that it becomes negative for some short period of time.

That $q(t)$ becomes negative means that the solution represents suction. During this suction period the cusp first becomes a branch point, which moves, and eventually this becomes a cusp again. At that stage the solution becomes univalent, and later on it joins the evolution to the right. This common evolution starts with $\Omega(t_4)$, but the time scale will be different since $q(t)$ has been modified. The above scenario is elaborated in Sect. 10.3, where the trajectory marked "Ullemar" in Fig. 10.2 represents the evolution described above. The two cusps appear when this trajectory intersects the line L_3, see Fig. 10.1.

1.3   On the Construction of a Branched Riemann Surface

The fact that the branch points B_j do not move in the case of the Löwner-Kufarev equation means that one can view the solution $f(\cdot, t)$ of this equation as a map of the unit disk into a fixed Riemann surface, as discussed above. The problem is that this Riemann surface is not known in advance, but has to be constructed along with the solution. Every time a zero of f' reaches $\partial\mathbb{D}$ the Riemann surface has to be updated with a new branch point, and then it has to be further extended to give space for the solution to develop. It is important in this process that $f(\cdot, t)$ is analytic in a disk $\mathbb{D}(0, \rho)$ which is strictly larger than the unit disk ($\rho > 1$), and then one has to make sure that this disk does not shrink. In fact, it actually increases, and this makes the process work out rather smoothly. These problems are dealt with in Chap. 5, specifically Lemma 5.2 for the increasing radius.

Next, having the extended Riemann surface the idea is to use the weak solution for the critical step when a zero of f' shall pass through $\partial\mathbb{D}$. The concept of a weak solution, for example formulated as partial balayage (1.6), can easily be adapted to branched Riemann surfaces (Chap. 4), and since a weak solution is unique it necessarily gives the correct solution to work with. However, here enters a new problem because the weak solution only consists of a family of domains, and in general these may become multiply connected. In the present case there is no doubt that the domain should stay simply connected for a short period of time after the transition of the zero, but to prove this rigorously has turned out to be difficult. Despite considerable efforts we have not succeeded, and therefore we leave corresponding statement Lemma/Conjecture 5.3 simply as a conjecture, which is taken as an assumption in the statement of the main result Theorem 5.1.

With this assumption in force the process of extending the solution to larger and larger time intervals goes on in small steps, with adding new branch points whenever new zeros of f' reach $\partial\mathbb{D}$ from outside. Eventually one ends up with an unlimited branched Riemann covering surface over the complex plane such that there exists a global (in time) simply connected weak solution spreading on this. Alternatively, without using the language of Riemann surfaces, one can view this solution simply as a global weak solution of the Löwner-Kufarev equation (1.2).

In Chap. 6 we the study in detail solutions f of the Polubarinova-Galin and Löwner-Kufarev equations in cases when f' is a rational function, with particular emphasis on what happens under phase transitions, i.e. when zeros of f' enter the unit disk. The property of f' being rational is preserved in time, but the structure of the rational function changes as zeros of f' pass through $\partial \mathbb{D}$. When one zero, ω_1, passes through $\partial \mathbb{D}$, say at the point $1 \in \partial \mathbb{D}$, then (in the generic case) two new zeros ω_2, ω_3 and one double pole ζ_1 are created, and these move out from $\mathbb{D}$. The pole is the mirror point of the zero that passed into $\mathbb{D}$: $\zeta_1 = 1/\bar{\omega}_1$. Pictorially, the process means that a new time dependent factor is hanged onto $f'(\zeta, t)$ according to

$$1 = \frac{(\zeta - 1)(\zeta - 1)}{(\zeta - 1)(\zeta - 1)} \mapsto \frac{(\zeta - \omega_2)(\zeta - \omega_3)}{(\zeta - \zeta_1)^2}.$$

This process is quite interesting and, as indicated in Example 4.3, it also connects to the theory of contractive divisors in Bergman space [48]. See also Theorems 6.1 and 6.2, with illustrations in Example 6.1 and Sect. 7.1.3. Here Eqs. (7.5), (7.6) and Fig. 7.1, give further illustrations.

In the case that f is a third order polynomial, as in (1.3), the phase change means that f transforms into a rational function of the form

$$f(\zeta) = \frac{b_1\zeta + b_2\zeta^2 + b_3\zeta^3 + b_4\zeta^4}{1 - \bar{\omega}_1\zeta},$$

with ω_1 as above. So this represents the change for f (or $\tilde{f}$) making a continued Löewner-Kufarev evolution possible after $\tilde{\Omega}(t_4)$ in Fig. 1.2. Compare also Theorem 10.1. Rational solutions, and related algebraic and Abelian domains, which are specific versions of quadrature domains [42], have been much discussed in the book [119]. See also [1], for solutions with more general versions of logarithmic singularities, and [14] for how quadrature domains come up in fluid dynamics in general.

Some further examples are given in Chap. 7. In Sect. 7.1 we first elaborate in detail several different solutions of the Polubarinova-Galin equation starting out from an cardioid. Quite surprisingly, one of these solutions can be driven backward in time, as a solution representing suction out of a cardioid. However, this is at the price of allowing a pole inside the unit disk, so the interpretation of this solution remains somewhat unclear. In Sect. 7.2 we still make an attempt to explain these matters in a more general setting.

1.4 Moment Coordinates and the String Equation

In Chaps. 8, 9, 10 we discuss the Polubarinova-Galin equation and Laplacian growth from the point of view of integrable systems and Hamiltonian mechanics, with eyes pointing towards quantum theory and string theory. This material is much inspired by developments starting in the 1990s by M. Mineev-Weinstein, P. Wiegmann, A. Zabrodin and others, in which conformal maps are parametrized by harmonic moments (see below). Some relevant papers are [17, 62, 63, 66, 74, 79, 112, 116, 124]. Our aim is to extend some of the results in these papers to the non-univalent case.

By choosing $h(z) = z^k$ in (1.4) one sees that the harmonic moments

$$M_k = \frac{1}{\pi} \int_{\Omega(t)} z^k \, dxdy, \quad k = 0, 1, 2, \ldots$$

are preserved in time for solutions of (1.1) and (1.2), with exception for the first moment, M_0, which increases linearly. These moments on the other hand provide a full set of local coordinates parametrizing simply connected domains with analytic boundary. From this point of view, the time derivative $\partial/\partial t$ can be identified (up to a factor) with the derivative $\partial/\partial M_0$, which is taken then with the other moments kept fixed.

Thus we can write

$$f(\zeta) = f(\zeta; M_0, M_1, M_2, \ldots)$$

for univalent and normalized analytic functions in $\mathbb{D}$. The above means that, without reference to any fluid dynamical problem, the Polubarinova-Galin equation (1.1) can be written on the beautiful and purely mathematical form

$$\{f, f^*\} = 1, \tag{1.7}$$

where $f^*(\zeta) = \overline{f(1/\bar{\zeta})}$ and where the Poisson bracket is defined in general by

$$\{f_1, f_2\} = \zeta \frac{\partial f_1}{\partial \zeta} \frac{\partial f_2}{\partial M_0} - \zeta \frac{\partial f_2}{\partial \zeta} \frac{\partial f_1}{\partial M_0}.$$

Equation (1.7) is called the string equation, and it is to hold on $\partial \mathbb{D}$. Then it automatically holds in a full neighborhood of $\partial \mathbb{D}$ since the left member is analytic. The question of independence and completeness of moments is discussed from the point of view of partial balayage in Sects. 3.3 and 3.4. See also [90].

When f is no longer univalent, more precisely when f' has zeros in $\mathbb{D}$, then the harmonic moments are no longer enough to parametrize f, and, as indicated earlier, suitable new parameters to add are the branch points. This gives the description

$$f(\zeta) = f(\zeta; M_0, M_1, M_2, \ldots; B_1, B_2, \ldots).$$

Now M_0 becomes the time variable for solutions of the Löwner-Kufarev equation (1.2) since such solutions are characterized (among solutions of (1.1)) by the B_j staying fixed. It follows that the string equation (1.7) still holds, with an extended meaning of the Poisson bracket.

In Chap. 9 we study more general Laplacian evolutions obtained by varying other moments, like M_k or B_j, instead of varying just M_0. We take the point of view of

Hamiltonian mechanics and, following ideas from [74, 124] (etc.) for the univalent case, we arrive at equations on the form

$$\frac{\partial f}{\partial M_k} = \{f, \mathcal{H}_k\}.$$

for certain Hamiltonian functions $\mathcal{H}_k$, which turn out to be polynomials in ζ^{-1}. See Theorems 9.1 and 9.2.

In Chap. 10 we provide rigorous proofs of the string equation (1.7) in cases of polynomial and rational mapping functions. In the polynomial case the string equation actually has a very specific meaning, referring to an exact formula, due to O. Kuznetsova and V. Tkachev [66, 116], for the Jacobi determinant between harmonic moments and the coefficients of the polynomial. This formula was conjectured by C. Ullemar [117], and somewhat in relation to this we consider in the final Sect. 10.4 a uniqueness question for quadrature domains in the light of the Hele-Shaw evolution discussed in track **(3)** at the end of Sect. 1.2 (the dashed arrow in Fig. 1.2).

1.5 Outlooks to Physics

Ordinary Laplacian growth, in the plane for example, has an immense number of applications in physics and other sciences. We do not give any list here but just refer to a large number of papers and books, partly summarized in [43]. Examples of more recent surveys are [3] and [113]. Laplacian growth on Riemann surfaces (without branch points) are discussed in for example [53, 111, 119].

Laplacian growth on branched Riemann surfaces is, as we know it so far, merely a mathematical construction which is difficult to apply directly to physics. In fact, a covering map with branch points cannot, when it is viewed as a vertical projection (the standard picture), be realized in physical three dimensional space since different sheets will have to cross each other along the branch cuts. Thus any physical model need to be abstract in some way.

Still one may get some inspiration by turning to cosmology. In some contexts it is desirable to have available canonical exhaustions of the universe, or space-like slices of it, by increasing families of domains which topologically are balls (hence simply connected in the two dimensional case).

Clearly Laplacian growth provides a good candidate for such exhaustions. It can be started from empty space and, as has been shown by H. Hedenmalm and S. Shimorin [53], the domains obtained will be simply connected, provided the space is hyperbolic (has non-positive Gaussian curvature). See Sect. 5.3 below for a short summary.

Branched Riemann surfaces over the complex plane are hyperbolic manifolds (although with somewhat singular and degenerate curvature, see Sect. 5.3), hence the above ideas apply to them. In Sect. 4.2 we consider an initially trivial example

(Example 4.2) with Laplacian growth starting from empty space and with source at the origin in a copy $\mathbb{C}_1$ of the complex plane. The result is of course just an increasing family of disks. Consider now this $\mathbb{C}_1$ as our "universe" and introduce next a "parallel universe" $\mathbb{C}_2$, attached to $\mathbb{C}_1$ via a "wormhole" at som point a. More precisely, $\mathbb{C}_1$ and $\mathbb{C}_2$ are actually attached via a whole branch cut from a to ∞.

Now, after the growing disks in $\mathbb{C}_1$ have reached the point a the evolution will partly continue on the other sheet $\mathbb{C}_2$, and eventually both sheets will be filled with a growing family of simply connected subdomains of the Riemann surface obtained. Hence the entire universe, with two parallel sheets, will be exhausted by topological balls in this two dimensional case, as desired.

1.6   Acknowledgements

First of all warm thanks go to Joakim Roos for several years of collaboration, which includes important inputs to the book. Joakim Roos has been crucial for the development of the theory of partial balayage, one of the basic tools in the book. In addition, and more specifically, Roos has contributed with numerical work and graphics in connection with examples in Chap. 7. In this respect Figs. 7.2 and 7.3 are informative, and confirms in a convincing way the somewhat remarkable fact that if one leaves the traditional scope of univalent conformal maps and allows certain multi-valued functions, then one can achieve something which looks like a suction solution for a cardioid.

The authors are also grateful to Michiaki Onodera for important information and discussions and to Andreas Minne for various kind of help. We also thank the referees for careful reading and for providing relevant and constructive criticism, which has led to improvements of the presentation.

Chapter 2
The Polubarinova-Galin and Löwner-Kufarev Equations

Abstract The basic definitions and properties related to the Polubarinova-Galin and the Löwner-Kufarev equations are set up. This includes discussing subordinations and liftings to Riemann surfaces.

2.1 Basic Set Up in the Univalent Case

The **Polubarinova-Galin equation** is the dynamical equation for the conformal map from the unit disk onto a domain in the complex plane representing the two-dimensional view of a blob of a viscous fluid, which grows or shrinks due to the presence of a source or sink at one point, chosen to be the origin. The type of flow in question, actually incompressible potential flow in the two dimensional picture, is traditionally called **Hele-Shaw flow** (see [43, 120] for historical accounts), and in recent time also **Laplacian growth**, referring to the moving boundary problem with injection or suction. The dynamical law for the growth of the fluid domain is that the boundary moves in the outward normal direction with speed proportional to the normal derivative of the Green function with pole at the origin. This Green function then represents the pressure of the fluid.

Let $\mathscr{O}_{\text{univ}}(\overline{\mathbb{D}})$ denote the set of (germs of) functions f which are holomorphic and univalent (one-to-one) in a neighborhood of the closed unit disk $\overline{\mathbb{D}}$ and which are normalized by $f(0) = 0$, $f'(0) > 0$. A smooth map $t \mapsto f(\cdot, t) \in \mathscr{O}_{\text{univ}}(\overline{\mathbb{D}})$ is a **(strong) solution** of the Polubarinova-Galin equation if it satisfies

$$\text{Re}\left[\dot{f}(\zeta, t)\overline{\zeta f'(\zeta, t)} \right] = q(t) \quad \text{for } \zeta \in \partial\mathbb{D}. \tag{2.1}$$

Here $q(t)$ is a real-valued function, which is given in advance and which represents the strength of the source/sink. Typically $q = \pm 1$, which corresponds to injection (plus sign) or suction (minus sign) at a rate 2π. Since the transformation $t \mapsto -t$ changes q to $-q$ in (2.1) it is enough to discuss one of the cases $q > 0$ and $q < 0$. In general we shall take $q > 0$. To increase flexibility we allow q to depend on time, and occasionally also to vanish.

© The Author(s), under exclusive license to Springer Nature Switzerland AG 2021
B. Gustafsson, Y.-L. Lin, *Laplacian Growth on Branched Riemann Surfaces*,
Lecture Notes in Mathematics 2287, https://doi.org/10.1007/978-3-030-69863-8_2

Equation (2.1) expresses that the image domains $\Omega(t) = f(\mathbb{D}, t)$ evolve in such a way that

$$\frac{d}{dt} \int_{\Omega(t)} h \, dm = 2\pi q(t) h(0) \tag{2.2}$$

for every function h which is harmonic in a neighborhood of $\overline{\Omega(t)}$, and where $dm = dx dy$ denotes Lebesgue measure in the plane. Equation (2.2) means exactly that the speed of the boundary $\partial\Omega(t)$ in the normal direction equals $q(t)$ times the normal derivative of the Green function of $\Omega(t)$ with a pole at $z = 0$, see (4.9) for some details in a Riemann surface setting. The equivalence between (2.1) and (2.2) follows from the general formula

$$\frac{d}{dt} \int_{\Omega(t)} \varphi \, dm = \int_{\partial\mathbb{D}} \varphi(f(\zeta, t)) \mathrm{Re}\left[\dot{f}(\zeta, t)\overline{\zeta f'(\zeta, t)}\right] d\theta \quad (\zeta = e^{i\theta}), \tag{2.3}$$

valid for any smooth evolution $t \mapsto f(\cdot, t) \in \mathscr{O}_{\mathrm{univ}}(\mathbb{D})$ and for any smooth test function φ in the complex plane. Compare Lemma 3.1 below.

On choosing $h(z) = z^k$, $k = 0, 1, 2, \ldots$ in (2.2) it follows that the **harmonic moments**

$$M_k(t) = \frac{1}{\pi} \int_{\Omega(t)} z^k dm(z) = \frac{1}{2\pi i} \int_{\mathbb{D}} f(\zeta, t)^k |f'(\zeta, t)|^2 \, d\bar{\zeta} d\zeta \tag{2.4}$$

are conserved quantities, except for the first one, which by (2.2) is related to $q(t)$ by $\frac{d}{dt} M_0(t) = 2q(t)$. Thus

$$M_0(t) = M_0(0) + 2Q(t), \tag{2.5}$$

where $Q(t)$ is the accumulated source up to time $t > 0$:

$$Q(t) = \int_0^t q(s) ds. \tag{2.6}$$

Occasionally we may use $Q(t)$ also for $t < 0$, in which case it is negative (if $q > 0$).

Within the class of smooth (or monotone) evolutions of simply connected domains, Laplacian growth is characterized by the preservation of the moments $M_1, M_2, \ldots$. This is a consequence of Theorem 10.13 and Corollary 10.14 in [97]. See also Theorem 6.2 and Corollary 6.3 in [31]. The above moment conservation was first observed in a seminal paper [88] by S. Richardson.

The moment conservation property is a kind of complete integrability, which is built into the concept of weak solution to be introduced in Chap. 3. See Definition 3.1, where the preservation of moments M_k (together with (2.5)) is replaced by a stronger property (in form of an inequality) in which the role of the test functions $h(z) = z^k$ is taken over by general subharmonic functions. Moment conservation also connects the subject of Laplacian growth to theories of integrable hierarchies, where the Polubarinova-Galin equation appears as a "string equation". This will be discussed in Chaps. 8, 9, 10, but we already here prepare by introducing a suitable **Poisson bracket** notation, which casts (2.1) on the form

$$\{f, f^*\}_t = 2q(t). \tag{2.7}$$

Here

$$f^*(\zeta) = \overline{f(1/\bar{\zeta})} \tag{2.8}$$

is the **holomorphic reflection** of f in $\partial\mathbb{D}$, and for $f_1 = f_1(\zeta, t)$, $f_2 = f_2(\zeta, t)$ analytic in a neighborhood of $\partial\mathbb{D}$ (with respect to ζ) and smooth in t the Poisson bracket is defined by

$$\{f_1, f_2\}_t = \zeta \frac{\partial f_1}{\partial \zeta} \frac{\partial f_2}{\partial t} - \zeta \frac{\partial f_2}{\partial \zeta} \frac{\partial f_1}{\partial t}. \tag{2.9}$$

It is immediately verified that (2.1) is equivalent to (2.7) holding identically in a neighborhood of $\partial\mathbb{D}$.

One may consider the Eq. (2.1) on different levels of generality. It is natural to keep the normalization $f(0) = 0$, $f'(0) > 0$, in fact the coupling to (2.2) depends on this, but (2.1) makes sense without any assumptions of univalence for f, thus for any $f \in \mathcal{O}_{\text{norm}}(\overline{\mathbb{D}})$ (see in general Glossary, at the end of the book, for notations), at least as long one makes sure that $q(t) = 0$ whenever a zero of f' appears on $\partial\mathbb{D}$. In the locally univalent case, $f \in \mathcal{O}_{\text{locu}}(\overline{\mathbb{D}})$, i.e. with $f' \neq 0$ on $\overline{\mathbb{D}}$, the mathematical treatment of (2.1) is exactly the same as in the 'physical' case $f \in \mathcal{O}_{\text{univ}}(\overline{\mathbb{D}})$. We shall then speak of a **locally univalent solution** of the Polubarinova-Galin equation.

When $f \in \mathcal{O}_{\text{locu}}(\overline{\mathbb{D}})$, then $\dot{f}/\zeta f' \in \mathcal{O}(\overline{\mathbb{D}})$ and Eq. (2.1) can be solved for $\dot{f}$ by dividing both sides with $|\zeta f'|^2$. The result is an equation which we shall refer to as the **Löwner-Kufarev equation**, namely

$$\dot{f}(\zeta, t) = \zeta f'(\zeta, t) P(\zeta, t) \quad (\zeta \in \mathbb{D}). \tag{2.10}$$

Here $P(\zeta, t)$ is that analytic function in $\mathbb{D}$ whose real part has boundary value $q(t)|f'(\zeta, t)|^{-2}$ and which is normalized by $\text{Im}\, P(0, t) = 0$. Explicitly $P(\zeta, t)$ is given as the Poisson integral

$$P(\zeta, t) = \frac{1}{2\pi i} \int_{\partial\mathbb{D}} \frac{q(t)}{|f'(z, t)|^2} \frac{z + \zeta}{z - \zeta} \frac{dz}{z} \quad (\zeta \in \mathbb{D}). \tag{2.11}$$

When $f \in \mathcal{O}_{\text{locu}}(\overline{\mathbb{D}})$ then $P \in \mathcal{O}(\overline{\mathbb{D}})$, in fact the right member of (2.10) extends analytically as far as f does (see [28] for a proof). We shall keep the notation $P = P(\zeta, t)$ also for the analytic extension of the Poisson integral beyond $\overline{\mathbb{D}}$.

As a general notation, we set

$$g(\zeta, t) = f'(\zeta, t). \tag{2.12}$$

The function g in fact turns out to be more fundamental than f itself. Of course, f can be recaptured from g by

$$f(z, t) = \int_0^z g(\zeta, t)d\zeta. \tag{2.13}$$

Part of this treatise will deal with the case that g is a rational function, or perhaps better to say, $g\,d\zeta$ is a rational differential, in other words an **Abelian differential** on the Riemann sphere. If g has residues then f will have logarithmic poles, besides ordinary poles. The terminology **Abelian domain** for the image domain $\Omega = f(\mathbb{D})$ has been used [119] when g is rational. Alternatively one may speak of Ω being a **quadrature domain** (see [42, 109] for the terminology and further references), which in the present case means that a finite **quadrature identity** of the kind

$$\int_\Omega h(z)dxdy = \sum_{j=1}^r c_j \int_{\gamma_j} h(z)dz + \sum_{j=0}^\ell \sum_{k=1}^{n_j-1} a_{jk}h^{(k-1)}(z_j) \tag{2.14}$$

holds for integrable analytic functions h in Ω. Here the z_j are fixed (i.e. independent of h) points in Ω, with specifically $z_0 = 0$, the c_j, a_{jk} are fixed coefficients, and the γ_j are arcs in Ω with end points among the z_j. In case all coefficients c_j vanish, Ω is called a "classical quadrature domain", or sometimes an **algebraic domain**. This occurs if f itself is a rational function, not only g. The above sorts of structure are stable under Hele-Shaw flow because, as is seen from (2.2), what happens under the evolution is only that the right member is augmented by the term $2\pi Q(t)h(0)$, where $Q(t)$ is the accumulated source up to time t, see (2.6).

When g is rational we shall write it on the form

$$g(\zeta, t) = b(t)\frac{\prod_{k=1}^m(\zeta - \omega_k(t))}{\prod_{j=1}^n(\zeta - \zeta_j(t))} = b(t)\frac{\prod_{i=1}^m(\zeta - \omega_i(t))}{\prod_{j=1}^\ell(\zeta - \zeta_j(t))^{n_j}}. \tag{2.15}$$

Here $m \geq n = \sum_{j=1}^\ell n_j$, $|\zeta_j| > 1$ and repetitions are allowed among the ω_k, ζ_j to account for multiple zeros and poles. Then, with the argument of $b(t)$ chosen so that $g(0, t) > 0$, $f \in \mathcal{O}_{\text{locu}}(\overline{\mathbb{D}})$ if and only if $|\omega_k| > 1$, $|\zeta_j| > 1$ for all k and j. The assumption $m \geq n$ means that $g\,d\zeta$, as a differential, has at least a double pole at infinity, which the Hele-Shaw evolution in any case will force it to have because the source/sink at the origin creates a pole of f at infinity.

The form (2.15) is stable in time, with the sole exception that when $m = n$ the pole of f at infinity may disappear at one moment of time (see [35], or Proposition 6.1 below). The rightmost member of (2.15) will be used when we need to be explicit about the orders of the poles. The convention then is that $\zeta_1, \ldots, \zeta_\ell$ are distinct and $n_j \geq 1$. Thus $n = \sum_{j=1}^{\ell} n_j$, and in the full sequence $\zeta_1, \ldots, \zeta_n$, the tail $\zeta_{\ell+1}, \ldots, \zeta_n$ will be repetitions of (some of) the $\zeta_1, \ldots, \zeta_\ell$ according to their orders. In Eqs. (2.14) and (2.15), ℓ and the n_j are the same.

One can easily express the Löwner-Kufarev equation (2.10) directly in terms of g. In fact, on writing P_g for the Poisson integral in (2.11), Eq. (2.10) together with (2.13) is equivalent to

$$\frac{\partial}{\partial t} \log g(\zeta, t) = \zeta P_g(\zeta, t) \frac{\partial}{\partial \zeta} \log g(\zeta, t) + \frac{\partial}{\partial \zeta} (\zeta P_g(\zeta, t)). \tag{2.16}$$

When g is rational, as in (2.15), then also $P_g(\zeta, t)$ will be a rational function (see more precisely (6.9) in Sect. 6.2), and so will the derivatives of $\log g$. We have

$$\log g(\zeta, t) = \log b(t) + \sum_{k=1}^{m} \log(\zeta - \omega_k(t)) - \sum_{j=1}^{n} \log(\zeta - \zeta_j(t)), \tag{2.17}$$

$$\frac{\partial}{\partial t} \log g(\zeta, t) = \frac{\dot{b}(t)}{b(t)} - \sum_{k=1}^{m} \frac{\dot{\omega}_k(t)}{\zeta - \omega_k(t)} + \sum_{j=1}^{n} \frac{\dot{\zeta}_j(t)}{\zeta - \zeta_j(t)}, \tag{2.18}$$

$$\frac{\partial}{\partial \zeta} \log g(\zeta, t) = \sum_{k=1}^{m} \frac{1}{\zeta - \omega_k(t)} - \sum_{j=1}^{n} \frac{1}{\zeta - \zeta_j(t)}. \tag{2.19}$$

Thus (2.16) becomes an identity between rational functions.

2.2 Dynamics and Subordination

In the non locally univalent case the Polubarinova-Galin and Löwner-Kufarev equations are no longer equivalent. The Löwner-Kufarev equation is the stronger one, and solutions to it can still be viewed as univalent mapping functions, but then onto subdomains of a Riemann surface. The evolution of these subdomains is monotone, which amounts to saying that the function family is a subordination chain. Solutions to the more general Polubarinova-Galin equation are not unique, and are monotone only in the sense that the counting function is monotone.

Definition 2.1 For any $f \in \mathcal{O}(\overline{\mathbb{D}})$, the **counting function**, or **mapping degree**, ν_f of f tells how many times a value $z \in \mathbb{C}$ is attained by f in $\mathbb{D}$. It is an integer valued function defined almost everywhere in $\mathbb{C}$ (namely outside $f(\partial \mathbb{D})$) by

$$\nu_f(z) = \text{card}\{\zeta \in \mathbb{D} : f(\zeta) = z\} = \frac{1}{2\pi i} \int_{\partial \mathbb{D}} d \log (f(\zeta) - z). \tag{2.20}$$

Clearly, f is univalent in $\mathbb{D}$ if and only if $0 \leq \nu_f \leq 1$.

Definition 2.2 Let $f, g \in \mathcal{O}_{\mathrm{norm}}(\mathbb{D})$. We say that f is **subordinate** to g, and write $f \prec g$, if there exists a univalent function $\varphi : \mathbb{D} \to \mathbb{D}$ such that $f = g \circ \varphi$. Note that φ is automatically normalized, hence $\varphi \in \mathcal{O}_{\mathrm{univ}}(\mathbb{D})$.

Let $I \subset \mathbb{R}$ be any interval. A map $I \ni t \mapsto f(\cdot, t) \in \mathcal{O}_{\mathrm{norm}}(\mathbb{D})$ is called a **subordination chain** on I if $f(\cdot, s) \prec f(\cdot, t)$ whenever $s \leq t$.

The following lemma shows that by increasing the level of abstraction (lifting the maps to a Riemann surface), subordination becomes nothing else than ordinary monotonicity. The result is not new (see [86]), but we give the proof because it will be a model for how the specific Riemann surfaces needed for the Hele-Shaw problem will be constructed.

Lemma 2.1 *Let* $\{f(\cdot, t)\}_{t \in I} \subset \mathcal{O}_{\mathrm{norm}}(\mathbb{D})$, *where* $I \subset \mathbb{R}$ *is any interval. Then the following are equivalent.*

(i) $\{f(\cdot, t)\}$ is a subordination chain on I.

(ii) There exists a Riemann surface $\mathcal{M}$, a nonconstant analytic function
$$p : \mathcal{M} \to \mathbb{C} \text{ ("covering map") and univalent analytic functions}$$

$$\tilde{f}(\cdot, t) : \mathbb{D} \to \mathcal{M} \quad (t \in I)$$

("liftings" of the $f(\cdot, t)$) such that

(a) $f(\zeta, t) = p(\tilde{f}(\zeta, t))$;

(b) $\tilde{f}(\mathbb{D}, s) \subset \tilde{f}(\mathbb{D}, t)$ for $s \leq t$.

Proof The proof that (ii) implies (i) is just a straight-forward verification, with the subordination functions defined by

$$\varphi(\zeta, s, t) = \tilde{f}^{-1}(\tilde{f}(\zeta, s), t) \tag{2.21}$$

for $s \leq t$, and where $\tilde{f}^{-1}(\zeta, t)$ denotes the inverse of $\tilde{f}(\zeta, t)$ with respect to ζ.

To prove the opposite, assume (i). We have to construct the Riemann surface $\mathcal{M}$ and the covering map p. For each $t \in I$, let $\mathbb{D}_t$ be a copy of $\mathbb{D}$ and let $\mathcal{M}_t$ be $\mathbb{D}_t$ considered as an abstract Riemann surface (for which $\mathbb{D}_t$ serves as coordinate space). We define a covering map

$$p(\cdot, t) : \mathcal{M}_t \to \mathbb{C}$$

by declaring that, in the coordinate space $\mathbb{D}_t$, it shall be represented by

$$f(\cdot, t) : \mathbb{D}_t \to \mathbb{C}.$$

When $s \leq t$ we have the embedding $\varphi(\cdot, s, t) : \mathbb{D}_s \to \mathbb{D}_t$ coming from the assumed subordination, which we on the level of the abstract Riemann surfaces consider as an inclusion map

$$\mathcal{M}_s \subset \mathcal{M}_t. \tag{2.22}$$

Note that these embeddings and inclusions commute with the covering maps because of the subordination relations

$$f(\varphi(\zeta, s, t), t) = f(\zeta, s) \quad (s \le t). \tag{2.23}$$

In view of the inclusions (2.22) we may define

$$\mathscr{M} = \cup_{t \in I} \mathscr{M}_t.$$

This is a Riemann surface because each point belongs to some $\mathscr{M}_t$, and there it has a neighborhood (e.g. all of $\mathscr{M}_t$) which can be identified with an open subset of the complex plane ($\mathscr{M}_t \cong \mathbb{D}_t \cong \mathbb{D}$), and the coordinates on $\mathscr{M}$ so obtained are related by invertible analytic functions (the $\varphi(\cdot, s, t)$). The covering map $p : \mathscr{M} \to \mathbb{C}$ is defined by declaring that on $\mathscr{M}_t$ it shall agree with $p(\cdot, t)$. Again, this is consistent.

Finally, the map $f(\cdot, t) : \mathbb{D} \to \mathbb{C}$ lifts to

$$\tilde{f}(\cdot, t) : \mathbb{D} \to \mathscr{M}_t \subset \mathscr{M}$$

by declaring that on identifying $\mathscr{M}_t$ with $\mathbb{D}_t$ it shall simply be the identity map: $\tilde{f}(\zeta, t) = \zeta \in \mathbb{D}_t$ for $\zeta \in \mathbb{D}$. Also this is consistent.

In the last picture, the evolution maps $\tilde{f}(\zeta, t)$ become trivial, while the covering maps are nontrivial ($p(\zeta, t) = f(\zeta, t)$):

$$\mathbb{D} \xrightarrow{\text{id}} \mathbb{D}_t \xrightarrow{f(\cdot, t)} \mathbb{C}.$$

For visualization it may however be better to have the view

$$\mathbb{D} \xrightarrow{\tilde{f}(\cdot, t)} \mathscr{M}_t \xrightarrow{\text{proj}} \mathbb{C}$$

in which the evolution maps $\tilde{f}(\zeta, t)$ really are liftings of the $f(\zeta, t)$, while the covering maps $p(\cdot, t)$ are trivial identifications (local identity maps, except at branch points).

Now, when $\mathscr{M}$ and p have been constructed the rest of the proof are easy verifications (omitted). □

Example 2.1 The functions

$$f(\zeta, t) = \frac{\zeta(t^3\zeta - 2t^2 + 1)}{\zeta - t}$$

can be shown to make up a non-univalent subordination family on the interval $1 < t < \infty$. The derivative $f'(\zeta, t)$ vanishes at $\zeta = t^{-1} \in \mathbb{D}$. The Riemann surface $\mathscr{M}$ appearing in Lemma 2.1 consists, when visualized as a covering surface over $\mathbb{C}$, of two copies of $\mathbb{C}$ joined by a branch point at $f(t^{-1}, t) = 1$. This example will be further discussed in Example 4.3, where also partial proofs of the above statements can be found.

Lemma 2.2 *For $f, g \in \mathscr{O}_{norm}(\mathbb{D})$, $f \prec g$ implies $v_f \leq v_g$ (almost everywhere).*

Proof This is immediate from a change of variable in the integral appearing in v_f: assuming $f = g \circ \varphi$ we have, using that φ is univalent,

$$v_f(z) = \frac{1}{2\pi i} \int_{\partial \mathbb{D}} d \log (f(\zeta) - z) = \frac{1}{2\pi i} \int_{\partial \mathbb{D}} d \log (g(\varphi(\zeta)) - z)$$

$$= \frac{1}{2\pi i} \int_{\varphi(\partial \mathbb{D})} d \log (g(\zeta) - z) \leq \frac{1}{2\pi i} \int_{\partial \mathbb{D}} d \log (g(\zeta) - z) = v_g(z).$$

$\square$

2.3 The Polubarinova-Galin Versus the Löwner-Kufarev Equation

The relationship between the Polubarinova-Galin and the Löwner-Kufarev equations in the non-univalent case is the following.

Theorem 2.1 *Let $I \ni t \mapsto f(\cdot, t) \in \mathscr{O}_{norm}(\overline{\mathbb{D}})$ be smooth on some time interval I and assume that $f' \neq 0$ on $\partial \mathbb{D}$ on this interval. Then for $q(t) \geq 0$ the following are equivalent.*

(i) $f(\zeta, t)$ solves the Löwner-Kufarev equation (2.10).
(ii) $f(\zeta, t)$ solves the Polubarinova-Galin equation (2.1) and $\dot{f}(\omega, t) = 0$ for every root $\omega \in \mathbb{D}$ of $f'(\omega, t) = 0$.
(iii) $f(\zeta, t)$ solves the Polubarinova-Galin equation (2.1) and $\{f(\cdot, t)\}$ is a subordination chain.

Remark 2.1 As a matter of terminology, if $\omega(t) \in \mathbb{D}$ is a zero of f', then $f(\omega(t), t)$ will be called a **branch point** of f, viewing f as a covering map. In a related terminology, it is branch point of the then multivalued lifting map f^{-1}. Since $\frac{d}{dt} f(\omega(t), t) = \dot{f}(\omega(t), t)$ when $f'(\omega(t), t) = 0$, the second condition in (ii) expresses that the branch points do not move.

Proof The additional condition in (ii) means more precisely (taking multiplicities into account) that

$$\frac{\dot{f}(\zeta, t)}{\zeta f'(\zeta, t)} \in \mathscr{O}(\overline{\mathbb{D}}). \tag{2.24}$$

After dividing both members in (2.1) by $|\zeta f'(\zeta, t)|^2$ and using the defining properties (2.11), (2.12) of $P(\zeta, t)$ this condition is seen to be exactly what is needed to pass between (2.1) and (2.10). Thus (i) and (ii) are equivalent.

Assume next that (iii) holds. That $\{f(\cdot, t)\}$ is a subordination chain means that for $s \leq t$ there exist univalent functions $\varphi(\cdot, s, t) : \mathbb{D} \to \mathbb{D}$ such that (2.23) holds. By differentiating (2.23) with respect to t it immediately follows that (2.24) holds. Thus (iii) implies (ii).

We finally prove that (i) implies (iii). This is done exactly as in the corresponding proof for Löwner chains of univalent functions in Chapter 6 of [86]. To construct the subordination functions $\varphi(\cdot, s, t)$ one considers, for given $s \geq 0$ and $\zeta \in \mathbb{D}$, the initial value problem

$$\begin{cases} \frac{dw}{dt} = -w P(w, t), & t \geq s, \\ w(s) = \zeta. \end{cases} \tag{2.25}$$

It has a unique solution $w = w(t)$ defined on the time interval on which f, and hence P, is defined. In terms of $w(t)$ we then define, for $s \leq t$,

$$\varphi(\zeta, s, t) = w(t).$$

Since different trajectories for (2.25) never intersect $\varphi(\zeta, s, t)$ is a univalent function of ζ in the unit disk, and using the chain rule and (2.10) one sees that $\frac{d}{dt} f(\varphi(\zeta, s, t), t) = 0$. Thus $f(\varphi(\zeta, s, t), t)$ is constantly equal to its initial value at $t = s$, which is $f(\zeta, s)$. This proves the subordination. $\qquad\square$

The Polubarinova-Galin equation itself is equivalent to a more general version of the Löwner-Kufarev equation, as follows.

Theorem 2.2 *Let $I \ni t \mapsto f(\cdot, t) \in \mathcal{O}_{norm}(\overline{\mathbb{D}})$ be smooth on some time interval I and assume that $f' \neq 0$ on $\partial\mathbb{D}$ on this interval. Then $f(\zeta, t)$ solves the Polubarinova-Galin equation (2.1) if and only if*

$$\dot{f}(\zeta, t) = \zeta f'(\zeta, t) \left(P(\zeta, t) + R(\zeta, t) \right), \tag{2.26}$$

where P is the Poisson integral (2.11) and where $R(\zeta, t)$ is any function of the form

$$R(\zeta, t) = -i \operatorname{Im} \sum_{\omega_j \in \mathbb{D}} \sum_{k=1}^{r_j} \frac{2 B_{jk}(t)}{(-\omega_j(t))^k} + \sum_{\omega_j \in \mathbb{D}} \sum_{k=1}^{r_j} \left(\frac{2 B_{jk}(t)}{(\zeta - \omega_j(t))^k} - \frac{2\overline{B_{jk}(t)}\zeta^k}{(1 - \overline{\omega_j(t)}\zeta)^k} \right). \tag{2.27}$$

Here $\{\omega_j\}$ are the zeros of f' in $\mathbb{D}$ (necessarily finitely many), r_j is the order of the zero at $\{\omega_j(t)\}$, and $B_{jk}(t)$ are arbitrary smooth functions of t.

Proof The proof of Theorem 2.2 is immediate since the additional term $R(\zeta, t)$ satisfies

$$\operatorname{Re} R(\zeta, t) = 0, \quad \zeta \in \partial \mathbb{D}, \tag{2.28}$$

$$\operatorname{Im} R(0, t) = 0$$

and is allowed to contain exactly those kinds of singularities in $\mathbb{D}$ which will be killed by the factor f' in front of it in (2.26). The first term in (2.27) is just the normalization assuring that $\operatorname{Im} R(0, t) = 0$, and the other terms exchange polar parts between $\mathbb{D}$ and $\mathbb{C} \setminus \overline{\mathbb{D}}$ without changing the real part on $\partial \mathbb{D}$.   □

Remark 2.2 The term $R(\zeta, t)$ primarily regulates the motion of the branch points, but this means that it, indirectly, also affects the dynamics of the boundary $\partial \Omega(t)$. If $\omega_j \in \mathbb{D}$ is a simple zero, for example, then it follows from (2.26), (2.27) that B_{j1} is proportional to the speed of the corresponding branch point:

$$\frac{d}{dt} f(\omega_j(t), t) = 2\omega_j(t) f''(\omega_j(t), t) B_{j1}(t). \tag{2.29}$$

As for the dynamics of the boundary, we first remark that when $\zeta \in \partial \mathbb{D}$ is kept fixed the point $z = f(\zeta, t) \in \partial \Omega(t)$ generally does not move perpendicular to the boundary (even if $R = 0$). On decomposing the speed into normal and tangential directions,

$$\dot{f} = \dot{f}_{\text{normal}} + \dot{f}_{\text{tangential}} = \operatorname{Re}(\dot{f} \cdot \frac{\overline{\zeta f'}}{|f'|}) \frac{\zeta f'}{|f'|} + \operatorname{Im}(\dot{f} \cdot \frac{\overline{\zeta f'}}{|f'|}) \frac{i \zeta f'}{|f'|},$$

one sees from (2.28) that, in the instantaneous picture, the term $R(\zeta, t)$ only affects the tangential component $\dot{f}_{\text{tangential}}$. However, the whole configuration changes by time and the dynamics of $\partial \Omega(t)$ will still be (indirectly) influenced by $R(\zeta, t)$. This will be seen in examples in Sect. 7.1. One may also consider (2.1) in the case that $q(t) = 0$ (identically), and then the dynamics will only be that the branch points move, a kind of internal variation of the domain. Compare discussions in Sect. 8.3.

Chapter 3
Weak Solutions and Balayage

Abstract Weak solutions, of variational inequality type, are introduced. Their defining properties can be equivalently expressed in terms of quadrature identities for subharmonic functions, or in terms of partial balayage. Some versions of inverse balayage are also discussed, this needed as a preparatory step for constructing more general Laplacian evolutions. Finally, the exponential transform and the elimination function are introduced.

3.1 Weak Formulation of the Polubarinova-Galin Equation

Some of our main results will be formulated in terms of variational inequality weak solutions, just called weak solutions for short, which are expressed in terms of time independent test functions which are subharmonic in the domains $\Omega(t)$. We shall need such weak solutions also on Riemann surfaces, and pulling back then the solutions to the unit disk means effectively that we are using time dependent coordinates on the Riemann surfaces. And in terms of these coordinates the original test functions, which were stationary, become time dependent. We start by taking care of the difficulties which arise by these circumstances.

First consider a general smooth test function in the complex plane. When it is pulled back to the unit disk via the mapping functions f it becomes time dependent, and the time and space derivatives will be coupled. Indeed, if $\Phi(z)$ is any smooth function in $\mathbb{C}$ then, by the chain rule, the composed function $\Psi(\zeta, t) = \Phi(f(\zeta, t))$, defined for $\zeta \in \overline{\mathbb{D}}$, satisfies

$$|f'(\zeta, t)|^2 \frac{\partial \Psi}{\partial t} = \dot{f}(\zeta, t)\overline{f'(\zeta, t)} \frac{\partial \Psi}{\partial \zeta} + \overline{\dot{f}(\zeta, t)}f'(\zeta, t)\frac{\partial \Psi}{\partial \bar{\zeta}}, \qquad (3.1)$$

as is easily verified. When working in $\mathbb{D}$ we shall need test functions Ψ which satisfy just (3.1) in itself, without necessarily being of the form $\Phi \circ f$ for some Φ.

B. Gustafsson, Y.-L. Lin, *Laplacian Growth on Branched Riemann Surfaces*,
Lecture Notes in Mathematics 2287, https://doi.org/10.1007/978-3-030-69863-8_3

Lemma 3.1 *For any smooth evolution $t \mapsto f \in \mathcal{O}_{norm}(\overline{\mathbb{D}})$ and any smooth function $\Psi(\zeta, t)$ which satisfies (3.1) we have*

$$\frac{d}{dt} \int_{\mathbb{D}} \Psi(\zeta, t) |f'(\zeta, t)|^2 \, dm(\zeta) = \int_{\partial \mathbb{D}} \Psi(\zeta, t) \mathrm{Re}\left[\dot{f}(\zeta, t)\overline{\zeta f'(\zeta, t)} \right] d\theta, \qquad (3.2)$$

where $\zeta = e^{i\theta}$ in the right member.

Proof Differentiation under the integral sign gives, using (3.1),

$$\frac{d}{dt} \int_{\mathbb{D}} \Psi(\zeta, t) |f'(\zeta, t)|^2 \, dm(\zeta) = \int_{\mathbb{D}} (\dot{f}\bar{f}'\frac{\partial \Psi}{\partial \zeta} + \dot{\bar{f}} f'\frac{\partial \Psi}{\partial \bar{\zeta}} + \Psi \dot{f}'\bar{f}' + \Psi f'\dot{\bar{f}}') \, dm$$

$$= \frac{1}{2i} \int_{\mathbb{D}} (\bar{f}'\frac{\partial}{\partial \zeta}(\dot{f}\Psi) + f'\frac{\partial}{\partial \bar{\zeta}}(\dot{\bar{f}}\Psi)) d\bar{\zeta}d\zeta$$

$$= \frac{1}{2i} \int_{\partial \mathbb{D}} \Psi(\dot{f}\bar{f}'d\zeta - \dot{\bar{f}} f'd\bar{\zeta})$$

$$= \int_{\partial \mathbb{D}} \Psi(\zeta, t) \mathrm{Re}\left[\dot{f}(\zeta, t)\overline{\zeta f'(\zeta, t)} \right] d\theta.$$

□

Corollary 3.1 *For any smooth evolution $t \mapsto f(\cdot, t) \in \mathcal{O}_{norm}(\overline{\mathbb{D}}))$ and any smooth function Φ in $\mathbb{C}$ we have*

$$\frac{d}{dt} \int_{\mathbb{C}} \Phi(z)v_{f(\cdot,t)}(z) dm(z) = \int_0^{2\pi} \Phi(f(\zeta, t)) \mathrm{Re}\left[\dot{f}(\zeta, t)\overline{\zeta f'(\zeta, t)} \right] d\theta. \qquad (3.3)$$

Proof Pulling the left member back to the unit disk by means of f gives

$$\frac{d}{dt} \int_{\mathbb{C}} \Phi(z)v_{f(\cdot,t)}(z) \, dm(z) = \frac{d}{dt} \int_{\mathbb{D}} \Phi(f(\zeta, t)) |f'(\zeta, t)|^2 \, dm(\zeta).$$

Since the composed function $\Psi(\zeta, t) = \Phi(f(\zeta, t))$ satisfies (3.1) the corollary follows immediately from Lemma 3.1.   □

Note that Corollary 3.1 is strictly weaker than Lemma 3.1. If for example $v_{f(\cdot,t)} = 2$ on some part of $\mathbb{C}$ then Lemma 3.1 allows the test function Ψ there to take different values on the two sheets of $f(\mathbb{D})$ lying above this part, which is not possible for the Φ in Corollary 3.1.

When $f(\zeta, t)$ solves the Polubarinova-Galin equation (2.1) we get

$$\frac{d}{dt} \int_{\mathbb{C}} \Phi(z)v_{f(\cdot,t)}(z) dm(z) = q(t) \int_0^{2\pi} \Phi(f(e^{i\theta}, t)) d\theta.$$

In particular, applying this to arbitrary $\Phi \geq 0$:

Corollary 3.2 *For any solution $t \mapsto f(\cdot, t) \in \mathcal{O}_{norm}(\mathbb{D})$ of the Polubarinova-Galin equation (2.1), with $q(t) \geq 0$, $\nu_{f(\cdot, t)}$ is an increasing function of t.*

Specializing (3.3), on the other hand, to subharmonic and harmonic test functions (which we then denote h) we obtain, in view of the mean-value properties satisfied by such functions:

Corollary 3.3 *Let $t \mapsto f(\cdot, t) \in \mathcal{O}_{norm}(\overline{\mathbb{D}})$ solve the Polubarinova-Galin equation (2.1) with $q(t) \geq 0$. Then*

$$\frac{d}{dt} \int_{\mathbb{C}} h\nu_{f(\cdot, t)} dm \geq 2\pi q(t) h(0)$$

for any h which is subharmonic in a neighborhood of $\operatorname{supp} \nu_f$. If h is harmonic, equality holds.

As a particular case we get the relevant version of moment conservation. Keeping the rightmost member of (2.4) as definition of the harmonic moments in the non-univalent case, so that

$$M_k(t) = \frac{1}{2\pi i} \int_{\mathbb{D}} f(\zeta, t)^k |f'(\zeta, t)|^2 dm(\zeta) = \frac{1}{\pi} \int_{\mathbb{C}} z^k \nu_{f(\cdot, t)}(z) dm(z),$$

we have

$$\frac{d}{dt} M_k(t) = 0, \quad k = 1, 2, 3, \ldots$$

under the assumptions of Corollary 3.3.

The essential step towards the concept of weak solution is integration of the dynamical quantities with respect to time. This is closely related to the **Baiocchi transform**, first used by C. Baiocchi [6–8] to transform certain problems in porous medium flow (the dam problem) into variational inequality formulations. In our case, because of the strong conservation properties (in principle complete integrability), integration with respect to time transforms the original dynamical problem into a series of elliptic problems, one for each instant of time.

Thus we integrate the inequality in Corollary 3.3 with respect to t, to obtain

Corollary 3.4 *Whenever $s \leq t$ and h is subharmonic in a neighborhood of $\operatorname{supp} \nu_f(\cdot, t)$ we have, when f solves the Polubarinova-Galin equation (2.1),*

$$\int_{\mathbb{C}} h\nu_{f(\cdot, t)} dm - \int_{\mathbb{C}} h\nu_{f(\cdot, s)} dm \geq 2\pi (Q(t) - Q(s)) h(0), \tag{3.4}$$

where Q is the accumulated source (see (2.6)).

Here (3.4) can be viewed as a **weak formulation of the Polubarinova-Galin equation** (2.1). In the univalent case, namely when $0 \leq v_f \leq 1$, the statement (3.4) becomes right away the standard definition of a weak solution:

Definition 3.1 With $I \subset \mathbb{R}$ an interval (of any sort), a family of bounded open sets $\{\Omega(t) \subset \mathbb{C} : t \in I\}$ is a **weak solution** of the planar Laplacian growth (or Hele-Shaw) problem with a point source at the origin if for any $s, t \in I$ with $s \leq t$, $\Omega(s) \subset \Omega(t)$ and

$$\int_{\Omega(t)} h \, dm - \int_{\Omega(s)} h \, dm \geq 2\pi(Q(t) - Q(s))h(0) \tag{3.5}$$

holds for every h which is subharmonic and integrable in $\Omega(t)$. If the interval I is of the form $[0, T)$ (or $[0, T]$) then it is enough that (3.5) holds for $s = 0$ to have the full strength of (3.5).

For later purposes we introduce a notation (due to Sakai [97]) for classes of subharmonic functions as above: if Ω is an open subset of a Riemann surface and $\lambda \geq 0$ is a measure on the Riemann surface (typically $\lambda = m$ if $\Omega \subset \mathbb{C}$) we set

$$SL^1(\Omega, \lambda) = \{s : \Omega \to \mathbb{R} \cup \{-\infty\} : s \text{ is subharmonic and } \int_{\Omega} |s| \, d\lambda < \infty\}.$$
$$\tag{3.6}$$

Given any initial bounded open set $\Omega(0)$, a weak solution in the sense of Definition 3.1 always exists on the interval $I = [0, \infty)$, and it is unique up to nullsets. If $\Omega(0)$ is connected and $0 \in \Omega(0)$, then also $\Omega(t)$ is connected for all $t > 0$. However, the domains $\Omega(t)$ need not be simply connected all the time, hence may be out of reach for the Polubarinova-Galin and Löwner-Kufarev equations, even though they do become simply connected for large enough $Q(t)$. See [43, 47] and references therein.

In addition to the above, weak solutions $\partial\Omega(t)$ always have real analytic boundaries $\partial\Omega(t)$. More precisely, starting with any (bounded) open set $\Omega(0)$, the set $\partial\Omega(t) \setminus \overline{\Omega(0)}$ has a real analytic defining function. Certain (analytic) singularities, essentially inward cusps and double points, may be present. Complete investigations of boundary regularity, and possible singular points, have been performed by M. Sakai [101, 103]. These investigations use the Schwarz function. In Sect. 3.5 we shall present an alternative approach, based on the exponential transform.

Regularity of weak solutions with respect to time is less clear. As will be seen in examples later on, the best one can hope for is good regularity (perhaps real analyticity) on time intervals where no changes of connectivity occur. But when topological changes occur only continuity with respect to time can be expected.

3.2 Weak Solutions in Terms of Balayage

The weak solution can be seen as an instance of a sweeping process called partial balayage, which under present circumstances results in quadrature domains for subharmonic functions. We formulate first this sweeping process in $\mathbb{C}$ and then adapt it to Riemannian manifolds according to our needs. Some general references are [26, 31, 40, 41, 89, 110].

The fixed data is a measure λ which (for the purpose of the present work) has a bounded density with respect to Lebesgue measure, i.e. satisfies (in terms of densities) $0 \leq \lambda \leq C$ for some constant C and, outside a compact set, is bounded from below:

$$\lambda \geq c > 0 \tag{3.7}$$

for some constant c.

Definition 3.2 With λ as above, let μ be a positive Radon measure with compact support in $\mathbb{C}$. Then **partial balayage** of μ to λ is defined as

$$\text{Bal}(\mu, \lambda) = \mu + \Delta u,$$

where u is the smallest locally integrable function satisfying

$$\begin{cases} u \geq 0, \\ \Delta u \leq \lambda - \mu. \end{cases} \tag{3.8}$$

Remark 3.1 (On Notation) Above Δu denotes the distributional Laplacian of u and (3.8) shall be interpreted as inequalities between distributions. The second inequality in (3.8) shows that Δu is actually a signed measure, and as such it could also be denoted $(\Delta u)m$, with Δu then referring to the density of the measure, in case this is absolutely continuous. However, Lebesgue measure m will usually not be written out in contexts like this, it will simply be identified with the distribution 1 (identically one).

Still m, or rather dm, will be written out in integrals. On writing $dm = dxdy$ it will sometimes be identified with the two-form $dx \wedge dy$, which means that it also is given an orientation.

The assumption (3.7) guarantees that there do exist functions u with compact support satisfying (3.8), and then it follows from general potential theory that a smallest such u exists, and also that it can be taken to be lower semicontinuous. In particular, the result $\text{Bal}(\mu, \lambda)$ of partial balayage will be a measure with compact support.

The requirements on u represent an obstacle problem on standard form, see [23, 60, 83] in general. The solution u can also be characterized by the requirement that the two inequalities in (3.8) shall hold in the complementary sense:

$$u(\lambda - \mu - \Delta u) = 0. \tag{3.9}$$

When (3.8) holds then (3.9) is equivalent to

$$\langle \lambda - \mu - \Delta u, u \rangle = 0,$$

where the left member is to be interpreted as the duality pairing between Sobolev spaces $H^{-1}(B_R)$ (for $\lambda - \mu - \Delta u$) and $H_0^1(B_R)$ (for u) for some sufficiently large ball B_R, and assuming in addition that $\mu \in H^{-1}(B_R)$, i.e. that μ has finite energy. The latter can always be achieved by mollifying with radially symmetric mollifiers. The ambient ball B_R should be so large that the solution $u \in H_0^1(B_R)$ of the obstacle problem has compact support in B_R, and then u (extended by zero outside B_R) will not depend on R.

In general, Bal (μ, λ) will be squeezed between two natural bounds,

$$\min\{\mu, \lambda\} \le \text{Bal}\,(\mu, \lambda) \le \lambda,$$

and the more detailed structure is (under our assumptions) that

$$\text{Bal}\,(\mu, \lambda) = \lambda \chi_\Omega + \mu \chi_{\mathbb{C} \setminus \Omega}. \tag{3.10}$$

Here Ω denotes the largest open set in which the equality case $\Delta u = \lambda - \mu$ holds in the sense of distributions. Expressed in another way:

$$\Omega = \mathbb{C} \setminus \text{supp}\,(\lambda - \text{Bal}\,(\mu, \lambda)).$$

It is called the **saturated set**, and it contains the noncoincidence set for the obstacle problem:

$$\{z \in \mathbb{C} : u(z) > 0\} \subset \Omega.$$

The inclusion may be strict, but under mild conditions the difference set is just a Lebesgue null-set.

Another general property of partial balayage is that it can be divided into smaller steps:

$$\text{Bal}\,(\mu_1 + \mu_2, \lambda_1) = \text{Bal}\,(\text{Bal}\,(\mu_1, \lambda_2) + \mu_2, \lambda_1) \tag{3.11}$$

provided $\lambda_1 \le \lambda_2 + \mu_2$. The latter assumption says that one is not allowed to sweep too much in the first step, necessary because there is no way to undo excessive balayage.

In view of (3.10), the saturated set Ω contains all information of the result of partial balayage. Another characterization of this set, directly in terms of μ and λ, is as follows:

$$\mu < \lambda \text{ on } \mathbb{C} \setminus \Omega, \tag{3.12}$$

$$\int_{\Omega} h \, d\mu \leq \int_{\Omega} h \, d\lambda \text{ for all } h \in SL^1(\Omega, \lambda). \tag{3.13}$$

Here (3.12) shall be interpreted as saying that $\mathbb{C} \setminus \Omega \subset \text{supp}\,((\lambda - \mu)_+)$, in other words that whenever $\mu \geq \lambda$ in some open set U it follows that $U \subset \Omega$.

The class $SL^1(\Omega, \lambda)$ (recall (3.6)) of test functions can, in (3.13), be replaced by just all logarithmic kernels $h(z) = \log |z - a|$ for $a \in \mathbb{C}$ together with all $h(z) = -\log |z - b|$ for $b \in \mathbb{C} \setminus \Omega$, see [97, 98]. With these test functions, (3.13) reduces to the statement that $u \geq 0$ in $\mathbb{C}$, $u = 0$ on $\mathbb{C} \setminus \Omega$, where u now denotes the logarithmic potential of $\mu \chi_\Omega - \lambda \chi_\Omega$ (so that $\Delta u = \lambda \chi_\Omega - \mu \chi_\Omega$). The proof of the equivalence between (3.10) and (3.12)–(3.13) then becomes straight-forward on noting in particular that the above u will be identical with the function u appearing in Definition 3.2.

We shall mostly consider $\text{Bal}\,(\mu, \lambda)$ in cases when there exists an open set $D \subset\subset \mathbb{C}$ such that $\mu \geq \lambda$ on D, $\mu = 0$ outside D. In such cases,

$$\text{Bal}\,(\mu, \lambda) = \lambda \chi_\Omega. \tag{3.14}$$

When $\lambda = m$, then (3.13) expresses that Ω is a **quadrature domain for subharmonic functions** for μ. This means that $\mu = 0$ outside Ω and that

$$\int_{\Omega} h \, d\mu \leq \int_{\Omega} h \, dm \tag{3.15}$$

holds for all $SL^1(\Omega, m)$, see [97] for detailed information.

In terms of partial balayage the defining property (3.5) of a weak solution $\Omega(t)$ takes the form

$$\text{Bal}\,(2\pi(Q(t) - Q(s))\delta_0 + \chi_{\Omega(s)}, 1) = \chi_{\Omega(t)} \quad (s < t),$$

where the function 1 represents Lebesgue measure.

Remark 3.2 Generally speaking, partial balayage destroys information: in for example (3.14), Ω is uniquely determined by μ, but many different measures μ give the same Ω. Therefore the balayage point of view, or the formulation with quadrature domains for subharmonic functions, embodies the fact that not only does Hele-Shaw flow preserve harmonic moments, so that the mass distributions $2\pi Q(t)\delta_0 + \chi_{\Omega(0)}$ and $\chi_{\Omega(t)}$ above are gravi-equivalent, but also that there is a time direction saying that the first mass distribution contains more information than the second. This reflects the fact that Laplacian growth is well-posed in one time direction (increasing t when $q > 0$) but ill-posed in the other. This is similar to properties of the heat equation, and also reminds of the role of entropy in statistical mechanics, which singles out one time direction.

Example 3.1 To illustrate the use of partial balayage, we note that the measure $\nu_{f(\cdot,t)}m$ in (3.4) may be swept to a measure of the form $\chi_{\Omega(t)}m$. On identifying m with 1 (see Remark 3.1) this statement becomes $\mathrm{Bal}\,(\nu_{f(\cdot,t)}, 1) = \chi_{\Omega(t)}$. This is the same as saying that $\int h\nu_{f(\cdot,t)}dm \leq \int_{\Omega(t)} h dm$ for $h \in SL^1(\Omega(t), m)$, as in (3.15) above. Taking $s = 0$ as initial time and assuming for simplicity that $f(\cdot, 0)$ is univalent, so that $\nu_{f(\cdot,0)} = \chi_{\Omega(0)}$ with $\Omega(0) = f(\mathbb{D}, 0)$, the inequality (3.4) gives

$$\int_{\Omega(t)} h dm - \int_{\Omega(0)} h dm \geq 2\pi\, Q(t)h(0)$$

for functions h subharmonic in $\Omega(t)$. In other words, $\{\Omega(t) : t \geq 0\}$ is the ordinary weak solution, possibly multiply connected, with initial domain $\Omega(0)$.

The evolution of $\nu_{f(\cdot,t)}$ can therefore be viewed as a refinement of the ordinary weak solution, a refinement in the sense that it contains more information. One can always pass from $\nu_{f(\cdot,t)}$ to $\chi_\Omega(t)$ by balayage, but there is in general no way to recover $\nu_{f(\cdot,t)}$ from $\chi_\Omega(t)$. An even more refined version of the evolution is obtained by lifting everything to a Riemann surface over $\mathbb{C}$, which will be discussed in Chap. 4.

Remark 3.3 Discrete and numerical versions of partial balayage can be traced back to the work of D. Zidarov [129]. Some closely related matters are the "smash sum" (adding two overlapping domains by sweeping the parts covered twice) [16] and probabilistic versions of (discrete) partial balayage by internal diffusion limited aggregation [68].

3.3 Inverse Balayage

Direct use of partial balayage can only give Laplacian growth in the well-posed direction, with swelling domains. However, by performing an initial step of inverse balayage one can start from a more favorable initial situation and from that create domain variations of more general type. The following theorem says that if a domain Ω has analytic boundary, then the mass distribution with unit density on Ω can be pushed into a compact subset of Ω without affecting its potential outside Ω. The statement is actually slightly stronger than so, namely the following.

Theorem 3.1 *Assume that $\Omega \subset \mathbb{C}$ has smooth analytic boundary. Then for $\varepsilon > 0$ sufficiently small there exists a compact subset $K \subset \Omega$ with smooth boundary such that balayage of the measure on K having uniform density $1 + \varepsilon$ gives Ω with unit density:*

$$\mathrm{Bal}\,((1 + \varepsilon)\chi_K, 1) = \chi_\Omega. \tag{3.16}$$

Proof Spelling out the meaning of (3.16) gives the following equivalent statement: There exists a compact set $K \subset \Omega$ and a function $u \in C^1(\overline{\Omega})$ such that

$$
\begin{cases}
(1+\varepsilon)\chi_K + \Delta u = \chi_\Omega & \text{in } \mathbb{C}, \\
u \geq 0 & \text{in } \mathbb{C}, \\
u = 0 & \text{in } \mathbb{C} \setminus \Omega.
\end{cases}
\tag{3.17}
$$

We shall construct this u by starting from outside and going inwards. It is a consequence of the Cauchy-Kovalevskaya theorem that in some neighborhood, inside Ω, of $\partial\Omega$ there exists a solution u to the Cauchy problem

$$
\begin{cases}
\Delta u = 1 & \text{in } \Omega, \text{ near } \partial\Omega, \\
u = |\nabla u| = 0 & \text{on } \partial\Omega.
\end{cases}
\tag{3.18}
$$

We extend this solution by zero outside Ω. The so obtained function u is continuously differentiable where it is defined, in particular there is no distributional contribution to Δu on $\partial\Omega$.

Near $\partial\Omega$, in Ω, $u(z)$ behaves essentially as $\frac{1}{2}d(z)^2$, where $d(z)$ denotes the distance to the boundary, in particular $u > 0$ there. Hence it is possible to find a level curve $u = \text{const} > 0$ such that u has a strictly positive slope when approaching that curve from outside. This curve is the boundary ∂K of a compact subset $K \subset \Omega$. Now extend u, which so far is defined in $\mathbb{C} \setminus K$, including ∂K, into int K in such a way that it is continuous across ∂K and solves $\Delta u = -\varepsilon$ in int K, for some constant $\varepsilon > 0$. Then u is defined in all $\mathbb{C}$, and becomes continuously differentiable everywhere except on ∂K. The constant $\varepsilon > 0$ is to be chosen so small that the distributional contribution to $-\Delta u$ on ∂K becomes a positive measure, call it σ. This is possible because of the strictly positive slope of u, from outside, on ∂K.

As a summary we now have

$$
\Delta u = \chi_{\Omega \setminus K} - \varepsilon\chi_K - \sigma,
$$

equivalently

$$
(1+\varepsilon)\chi_K + \sigma + \Delta u = \chi_\Omega.
$$

In addition $u = 0$ outside Ω and $u \geq 0$ everywhere. This shows that

$$
\text{Bal}\,((1+\varepsilon)\chi_K + \sigma, 1) = \chi_\Omega.
$$

It remains to get rid of σ, but this is easily done by an application of (3.11). Indeed, just balayage $(1+\varepsilon)\chi_K + \sigma$ to level $1 + \varepsilon$. This will make K swell a little, say to a domain D with $K \subset D \subset \overline{D} \subset \Omega$. Then

$$
\text{Bal}\,((1+\varepsilon)\chi_K + \sigma, 1+\varepsilon) = (1+\varepsilon)\chi_D.
$$

It follows from general regularity theory that ∂D is analytic, in particular has Lebesgue measure zero, hence D above can be replaced by $\overline{D}$, and this will be the K in the theorem. An application of (3.11) gives finally

$$\text{Bal}\,((1+\varepsilon)\chi_D, 1) = \text{Bal}\,(\text{Bal}\,((1+\varepsilon)\chi_K + \sigma, 1+\varepsilon), 1)$$
$$= \text{Bal}\,((1+\varepsilon)\chi_K + \sigma, 1) = \chi_\Omega,$$

as desired.   □

Remark 3.4 The system (3.18) has been presented as a one-sided Cauchy problem, with u defined to be zero outside Ω. One may also consider the corresponding two-sided Cauchy problem, by keeping $\Delta u = 1$ in a full neighborhood of $\partial\Omega$. With this choice of u, the function

$$S(z) = \bar{z} - 4\frac{\partial u}{\partial z} \tag{3.19}$$

is the **Schwarz function** [15, 109] of $\partial\Omega$. It is characterized by being analytic in a neighborhood of $\partial\Omega$ and satisfying

$$S(z) = \bar{z} \quad \text{for } z \in \partial\Omega.$$

The map $z \mapsto \overline{S(z)}$ is the anti-conformal reflection in $\partial\Omega$, and it is related to the conformal map $f : \mathbb{D} \to \Omega$ by

$$f^* = S \circ f. \tag{3.20}$$

If u is kept, as in (3.18), as a one-sided solution of the Cauchy problem, then the function $S(z)$ is a one-sided Schwarz function, which may exist even if $\partial\Omega$ has certain singularities. The relation between the regularity of $\partial\Omega$ and the existence of a **one-sided Schwarz function** is studied in depth in [101].

3.4 More General Laplacian Evolutions

Let μ be a distribution with compact support in a given domain $\Omega \subset \mathbb{C}$. By a **Laplacian evolution** generated by μ we mean a smooth family $\{\Omega(t) : -\varepsilon < t < \varepsilon\}$, with ε sufficiently small and $\Omega(0) = \Omega$, such that $\partial\Omega(t)$ moves with normal velocity equal to $\partial p/\partial n$, where, for any fixed t, $p = p(z, t)$ denotes the solution of the Dirichlet problem

$$\begin{cases} -\Delta p = \mu & \text{in } \Omega(t), \\ \quad p = 0 & \text{on } \partial\Omega(t). \end{cases} \tag{3.21}$$

This can be expressed, in analogy with (2.2), by saying that

$$\frac{d}{dt} \int_{\Omega(t)} h \, dm = \langle \mu, h \rangle \tag{3.22}$$

shall hold for every function h which is harmonic in a neighborhood of $\overline{\Omega(t)}$. The right member in (3.22) denotes the distributional action of μ on h. Integrating (3.22) with respect to t gives that

$$\int_{\Omega(t)} h \, dm = \int_{\Omega(0)} h \, dm + t \langle \mu, h \rangle \tag{3.23}$$

for all h which are harmonic in the closure of the union of all domains $\Omega(t)$ involved.

If μ is not a positive measure and moreover $t > 0$, such an evolution can exist only if the boundary $\partial \Omega(0)$ is analytic and $\varepsilon > 0$ is sufficiently small. The main message for this section is that, on the other hand, assuming the analyticity of $\partial \Omega(0)$, such a local Laplacian evolution does exist and can be constructed by partial balayage. As a first step one has to convolve μ with a positive and radially symmetric mollifier with small support. This does not affect its action on harmonic functions, in (3.22) and (3.23) for example, so we may as well assume from beginning that μ is represented by a smooth function, call it ρ. Thus

$$\langle \mu, h \rangle = \int h \rho \, dm$$

for h harmonic in a neighborhood of the support of ρ.

The next step is to represent Ω by a compactly supported distribution as in Theorem 3.1. The point here is that the added level of density, from 1 to $1 + \varepsilon$, gives a margin within which one can make modifications, namely by $t \rho$ for $|t|$ sufficiently small.

Theorem 3.2 Let Ω be a domain with smooth analytic boundary, μ a distribution with compact support in Ω. Then there exists a corresponding Laplacian evolution $\Omega(t)$, defined for $|t|$ sufficiently small, such that (3.23) holds. With ρ a mollified version of μ as above and Ω represented as in (3.16), where K, ρ and ε are to be adapted so that $\operatorname{supp} \rho \subset K$, the evolution is given by

$$Bal\left((1 + \varepsilon) \chi_K + t \rho, 1\right) = \chi_{\Omega(t)}, \tag{3.24}$$

for $|t|$ so small that $|t| \sup |\rho| < \varepsilon$.

Proof Under the stated assumption on $|t|$ we have that $(1 + \varepsilon) \chi_K + t \rho \geq 1$ on K, and then it follows from general properties of partial balayage that the result of the balayage in (3.24) is of the form in the right member for some open set $\Omega(t)$. The domain $\Omega(t)$ will even have analytic boundary. Implicit in the balayage is that the result of integrating harmonic functions is the same after balayage as before. Using this for (3.16) and (3.24) gives

$$\int_{\Omega(t)} h \, dm = \int h((1+\varepsilon)\chi_K + t\rho) \, dm = \int_\Omega h \, dm + t\langle \mu, h \rangle,$$

as desired.   □

The theorem does not assert any smoothness with respect to t. The solution is indeed smooth in t, but the proof of this is more difficult and uses different methods, like working with the corresponding differential equations (in t) and making careful estimates in formulations with scales of Banach spaces. See [87], for example. An alternative approach is working with the corresponding obstacle problems and obtaining stability estimates, as in [107, 108].

Writing (3.23) (or (3.24)) in terms of the potential u in (3.17) it becomes

$$\chi_{\Omega(t)} = \chi_{\Omega(0)} + t\mu + \Delta u.$$

Here $u = u(t)$ and $\partial\Omega(t)$ depend smoothly on t, as indicated above, hence we can differentiate to obtain

$$\frac{\partial}{\partial t}\chi_{\Omega(t)} = \mu + \Delta\frac{\partial u}{\partial t}.$$

Thus

$$\begin{cases} -\Delta\frac{\partial u}{\partial t} = \mu & \text{in } \Omega(t), \\ \frac{\partial u}{\partial t} = 0 & \text{on } \partial\Omega(t), \end{cases}$$

the latter because u vanishes to the second order on $\partial\Omega(t)$ by (3.18). This is the same system as that satisfied by p above (see (3.21)), hence we conclude that

$$p = \frac{\partial u}{\partial t}, \tag{3.25}$$

or, in the other direction,

$$u(z, t) = \int_0^t p(z, \tau) \, d\tau.$$

The step $p \mapsto u$ is the previously mentioned Baiocchi transform.

Example 3.2 Choose

$$\mu = \frac{(-1)^k}{k!}\frac{\partial^k}{\partial x^k}\delta_0 = \frac{(-1)^k}{k!}\left(\frac{\partial}{\partial z} + \frac{\partial}{\partial \bar{z}}\right)^k\delta_0,$$

where δ_0 is the Dirac measure at $0 \in \Omega$. For $h(z) = z^n$, $n = 0, 1, 2, \ldots$ (clearly we can allow complex-valued harmonic functions in (3.23)) this gives

$$\langle \mu, h \rangle = \frac{(-1)^k}{k!} \langle (\frac{\partial}{\partial z} + \frac{\partial}{\partial \bar{z}})^k \delta_0, z^n \rangle - \frac{1}{k!} \langle \delta_0, (\frac{\partial}{\partial z} + \frac{\partial}{\partial \bar{z}})^k z^n \rangle$$

$$= \begin{cases} 0 & \text{if } k \neq n, \\ 1 & \text{if } k = n. \end{cases}$$

It follows that under the Laplacian evolution defined by μ all harmonic moments except M_k are conserved, while M_k changes linearly in time. So by Theorem 3.2 we have constructed such an evolution by means of partial balayage.

The above example shows that the moments $M_0, M_1, M_2, \ldots$ are independent of each other. They are also a complete set of variables for determination of a simply connected domain, at least with respect to small variations of the boundary. This is a well-known fact, but let us give an independent proof. The idea of the result goes back to [97], and the proof is the same as that of Corollary 3.10 in [30]. For the formulation we need a slightly stronger version of simple connectedness: we call a domain (or open set) D **solid** if $D = \text{int} (\overline{D})$ and $\mathbb{C} \setminus \overline{D}$ is connected.

Theorem 3.3 *Let $\Omega \subset \mathbb{C}$ be a simply connected domain with analytic boundary. Then there exists a compact subset $K \subset \Omega$ such that no other solid domain D satisfying $K \subset D$ has the same harmonic moments as Ω.*

Remark 3.5 The analyticity of $\partial\Omega$ seems to be a crucial requirement, as well as the local nature of the statement, implicit in the requirement $K \subset D$. There are examples of different solid domains with piecewise smooth boundaries having the same harmonic moments, see for example [94, 99, 128].

Proof By Theorem 3.1 we know that there exists a measure μ with compact support K in Ω such that

$$\text{Bal} (\mu, 1) = \chi_\Omega. \tag{3.26}$$

The exact form of μ as given in Theorem 3.1 is however not important in the present proof.

Let U_μ denote the **logarithmic potential** of μ (normalized for the present purpose so that $-\Delta U_\mu = \mu$) and similarly U_Ω and U_D those of χ_Ω and χ_D. Then $\text{Bal} (\mu, 1) = \mu + \Delta u$, where $u = U_\mu - U_\Omega$. Assuming that Ω and D have the same harmonic moments it follows that $U_\Omega = U_D$ in some neighborhood of infinity. Since $U_\mu - U_\Omega = u = 0$ outside Ω, $u \geq 0$ in all $\mathbb{C}$, and it also follows $U_\mu - U_D = 0$ everywhere outside D.

Now set

$$w = U_D - U_\Omega = (U_\mu - U_\Omega) - (U_\mu - U_D) = u - (U_\mu - U_D).$$

On the closed set $\overline{\Omega} \setminus D$ we have $w = u \geq 0$. In the complementary (open) set $\mathbb{C} \setminus (\overline{\Omega} \setminus D) = (\mathbb{C} \setminus \overline{\Omega}) \cup D$, w is superharmonic. In fact, w is superharmonic in the full complement of $\overline{\Omega}$. It follows that $w \geq 0$ everywhere.

Next we claim that $\partial D \subset \overline{\Omega}$. Indeed, in the contrary case there would exist a point $z \in \partial D \setminus \overline{\Omega}$, and in a neighborhood of that point w is superharmonic, and not harmonic (since $-\Delta w = \chi_D - \chi_\Omega = \chi_D$ near z). Combining this with the fact that $w \geq 0$ everywhere and $w(z) = 0$ we arrive at the desired contradiction.

Now, $\partial D \subset \overline{\Omega}$ together with the topological and regularity assumptions on Ω and D implies that $D = \Omega$. $\square$

Example 3.3 A classical example, which illustrates the above theorem, is that of a disk, say $\Omega = \mathbb{D}$. Then one can take (minimal choice) $K = \{0\}$ and $\mu = \pi \delta_0$. The conclusion of the theorem is that the unit disk is the only solid domain containing the origin and having the moments $M_0 = 1$, $M_1 = M_2 = \cdots = 0$. The counterexample in [99] exactly refers to this case: there do exist plenty of simply connected (but nonsolid) domains with piecewise smooth boundary with the same moments as the disk. And they can be easily constructed by partial balayage, as is actually done in [99].

3.5 Regularity of the Boundary via the Exponential Transform

As indicated above (in Sect. 3.3) boundary regularity of weak solutions can be discussed via the Schwarz function, as in [101–103], or via general regularity theory for solutions of obstacle problems, see [83] for example. Recall (3.19), which relates the two functions involved to each other. Below we shall briefly discuss how boundary regularity of weak solutions can also be obtained by using the exponential transform. The exponential transform is the exponential of a double Cauchy transform, and it originally came up within operator theory, more exactly in determinantal formulas for hyponormal operators, see [38, 77, 84, 125]. A few specific references for the material below are [37, 45, 46].

Definition 3.3 Given a bounded weight function $\rho \geq 0$ with compact support we define its **exponential transform** as

$$E_\rho(z, w) = \exp[-\frac{1}{\pi} \int_{\mathbb{C}} \frac{\rho(\zeta) dm(\zeta)}{(\zeta - z)(\bar{\zeta} - \bar{w})}], \quad z, w \in \mathbb{C}.$$

When $z = w$ the integral above may diverge (to $+\infty$), and if so happens we agree that $E_\rho(z, z) = 0$.

The classical case is that $\rho = \chi_\Omega$ for some domain $\Omega \subset \mathbb{C}$, but we shall need to use the exponential transform also for multi-sheeted domains, hence with ρ integer valued and typically of the form $\rho = \nu_{f(\cdot, t)}$, referring here to (2.20), (3.4).

Clearly $E_\rho(z, w)$ is analytic in z, antianalytic in w, when the variable in question is outside the support of ρ. Let $F_\rho(z, w)$ denote the analytic/anti-analytic germ of $E_\rho(z, w)$ at infinity, as well as analytic continuations of that germ. On expanding $F_\rho(z, w)$ in a power series at infinity one sees that the germ $F_\rho(z, w)$ contains information of all complex moments $\int z^k \bar{z}^j \rho(z) dm(z)$ of ρ, hence it determines ρ completely (as a distribution). Within the support of ρ, $E_\rho(z, w)$ is no longer analytic. On the other hand, as seen by a direct calculation, the function

$$H_\rho(z, w) = \frac{E_\rho(z, w)}{(\bar{z} - \bar{w})^{\rho(z)}(z - w)^{\rho(w)}}$$

is analytic/antianalytic on pieces where ρ takes constant integer values.

Now, considering the case of a single-sheeted domain $\rho = \chi_\Omega$ to begin with, the crucial property of $E_\rho(z, w)$ for our considerations is that when $\Omega = \Omega(t)$ comes from a weak solution started at an earlier instant $s < t$, as in (3.5), then the germ $F_\rho(z, w)$ has an analytic/antianalytic continuation all the way to $\overline{\Omega(s)}$, a compact subset of $\Omega(t)$. And this analytic continuation restricted to the diagonal gives a real analytic defining function for $\partial\Omega(t)$ as

$$\partial\Omega : \quad F_\rho(z, z) = 0. \tag{3.27}$$

Note next that $E_\rho(z, w) = F_\rho(z, w)$ outside supp $\rho = \overline{\Omega}$ by analytic continuation; this is true also if $\mathbb{C} \setminus \overline{\Omega}$ has bounded components, even though it is less obvious in that case. Since $E_\rho(z, z) = 0$ for $z \in \Omega$ and for most points $z \in \partial\Omega$, even with nonzero gradient (there may be exceptional boundary points, however), (3.27) indeed is a good defining equation for $\partial\Omega$. Full details on the above matters are given in [37, 39].

The analytic continuation $F_\rho(z, w)$ of the germ of $E_\rho(z, w)$ can be made fully explicit in terms of the Schwarz function $S(z)$ of $\partial\Omega$ by

$$F_\rho(z, w) = \frac{(S(z) - \bar{w})(z - \overline{S(w)})}{(\bar{z} - \bar{w})(z - w)} E_\rho(z, w)$$

for z, w in Ω, as far inside $\partial\Omega$ as the Schwarz function is defined. This equation can also be written, and extended to the multi-sheeted case, i.e. with f non-univalent, as

$$F_\rho(z, w) = \left(\frac{S(z) - \bar{w}}{\bar{z} - \bar{w}}\right)^{\rho(z)} \left(\frac{z - \overline{S(w)}}{z - w}\right)^{\rho(w)} E_\rho(z, w). \tag{3.28}$$

Here the right member need to be appropriately interpreted since the Schwarz function has several branches. In general, only combinations of the Schwarz function such as

$$\left(\frac{S(z) - \bar{w}}{\bar{z} - \bar{w}}\right)^{\rho(z)}$$

can be given a good enough mening. For details on the meaning of (3.28) in the algebraic case (see (3.34) below) we refer to [46].

3.6   The Resultant and the Elimination Function

In case Ω is the conformal image of $\mathbb{D}$ under a rational function f, the exponential transform of Ω introduced above is closely related to what we call the elimination function between f and its holomorphically reflected map f^*. This is the case also if f is not univalent, and the relationship extends to the case that f is a meromorphic function on any compact symmetric Riemann surface. We briefly review these matters below, referring to [45, 46] for details.

We first have to define the meromorphic resultant $\mathscr{R}(g, h)$ between two meromorphic functions, g and h, on a compact Riemann surface.

Definition 3.4 Let g and h be two meromorphic functions on a compact Riemann surface $\mathscr{M}$. Then their **meromorphic resultant** is defined as the multiplicative action of h on the divisor of g (or vice versa):

$$\mathscr{R}(g, h) = h((g)). \tag{3.29}$$

Here (g) denotes the divisor of zeros (counted positive) and poles (counted negative) of g.

The meaning of (3.29) is best explained by a simple example.

Example 3.4 Let the Riemann surface be the Riemann sphere ($\mathscr{M} = \mathbb{P}$) and let

$$g(\zeta) = \frac{(\zeta - a)(\zeta - b)^3}{(\zeta - c)^2}$$

for some distinct points $a, b, c \in \mathbb{C}$. Then the divisor of g is (in additive notation)

$$(g) = 1 \cdot (a) + 3 \cdot (b) - 2 \cdot (c) - 2 \cdot (\infty).$$

Thus (without specifying h),

$$\mathscr{R}(g, h) = \frac{h(a)h(b)^3}{h(c)^2 h(\infty)^2}.$$

This makes sense for generic choices of h, even though there are exceptions (for example, there may appear expression ∞/∞ or $0/0$).

When g, h are rational it is obvious that $\mathscr{R}(g, h) = \mathscr{R}(h, g)$. This symmetry still holds in the general case, but it is then less trivial. In fact, it is equivalent to the Weil reciprocity theorem [27, 123]. See further [45]. The most important property of the resultant is that it vanishes if and only if g and h have a common zero or pole.

In terms of the meromorphic resultant one can define an elimination function as follows.

Definition 3.5 The **elimination function** of any two meromorphic functions, g and h, on a compact Riemann surface is

$$\mathscr{E}(z, w) = \mathscr{E}_{g,h}(z, w) = \mathscr{R}(g - z, h - w), \quad z, w \in \mathbb{C}. \tag{3.30}$$

The elimination function is always a rational function, more precisely of the form

$$\mathscr{E}_{g,h}(z, w) = \frac{Q(z, w)}{P(z)R(w)} \tag{3.31}$$

for some polynomials Q, P, R. If $z = g(\zeta)$, $w = h(\zeta)$ for some $\zeta \in \mathscr{M}$ then the functions $g - z$ and $h - w$ have a common zero (namely ζ), hence $\mathscr{E}_{g,h}(z, w) = 0$ by properties of resultants. It follows that

$$\mathscr{E}_{g,h}(g(\zeta), h(\zeta)) = 0 \quad (\zeta \in \mathscr{M})$$

holds identically. In particular one arrives at the well known fact [20] that any two meromorphic functions on a compact Riemann surface are polynomially related: in the notation of (3.31) we have $Q(g, h) = 0$.

Let now $f : \mathbb{D} \to \Omega$ be a rational conformal map. In case f is univalent the basic relationship between the elimination function for the pair f and f^* and the exponential transform of $\rho = \chi_\Omega$ is that they have the same germ at infinity. See [45], Theorem 7 there, for details and proof. This quite remarkable identity remains true also when f is not univalent:

Theorem 3.4 *Let* $f \in \mathscr{O}_{norm}(\overline{\mathbb{D}})$ *be a (non-constant) rational function and let*

$$\rho(z) = \nu_f(z)$$

be its counting function (see Definition 2.1). Then

$$E_\rho(z, w) = \mathscr{E}_{f,f^*}(z, \bar{w}) \quad (|z|, |w| >> 1),$$

The theorem is a special case of Theorem 2 in [46], and we refer to that publication for the proof.

In the previous section we introduced the notation $F_\rho(z, w)$ for the germ at infinity of $E_\rho(z, w)$. Thus, by Theorem 3.4,

$$F_\rho(z, w) = \mathscr{E}_{f,f^*}(z, \bar{w}) \quad (|z|, |w| >> 1), \tag{3.32}$$

and this relation persists to hold throughout the domains of analyticity of the two members. The rational expression (3.31) for $\mathscr{E}(z, w)$ is more exactly

$$\mathscr{E}_{f,f^*}(z, \bar{w}) = \frac{Q(z, \bar{w})}{P(z)\overline{P(w)}}, \tag{3.33}$$

where $P(z)$ is the monic polynomial with zeros at the image points (under f) of the poles of f^*, equivalently at the quadrature nodes z_j in the quadrature identity (2.14). Such an identity holds with vanishing coefficients c_j when f is rational function. The polynomial Q describes the boundary of Ω as

$$\partial \Omega : \quad Q(z, \bar{z}) = 0.$$

The relations (3.32), (3.33) still hold when f is no longer univalent (but still rational), and according to (3.28) they make the structure of $E_\rho(z, w)$ explicit as

$$E_\rho(z, w) = \left(\frac{\bar{z} - \bar{w}}{S(z) - \bar{w}}\right)^{\rho(z)} \left(\frac{z - w}{z - \overline{S(w)}}\right)^{\rho(w)} \mathscr{E}_{f,f^*}(z, w). \tag{3.34}$$

We refer to [45, 46] for proofs of the above statements, and further clarifications.

Chapter 4
Weak and Strong Solutions on Riemann Surfaces

Abstract Laplacian growth and partial balayage are extended to Riemann surfaces. The Riemann surfaces considered are more precisely covering surfaces of the complex plane, and the somewhat abstract theory is illustrated by examples. It is also shown that the partial balayage, which represents time evolution, commutes with the covering projection in a natural sense.

4.1 Laplacian Growth on Manifolds

Laplacian growth makes sense on Riemannian manifolds (of any dimension). The only difference compared to the Euclidean case then is that the measure $dm = dx \wedge dy$ in, for example, (2.2) and (3.5) shall be replaced by the intrinsic volume form of the manifold. This also indicates how (3.5) changes under variable transformations: $dm = dx \wedge dy$ shall be treated as a 2-form. We shall need to make these things precise in the case that the Riemannian manifold is a branched covering Riemann surface over $\mathbb{C}$, with the metric inherited from the Euclidean metric on $\mathbb{C}$ via the covering map.

Let $\mathscr{M}$ be a Riemann surface and $p : \mathscr{M} \to \mathbb{C}$ a non-constant analytic function, thought of as a, possibly branched, covering map. If $\tilde{z} = \tilde{x} + i\tilde{y}$ is a local holomorphic coordinate on $\mathscr{M}$ and $z = x + iy$ the usual coordinate on $\mathbb{C}$ then the Riemannian metric on $\mathscr{M}$ is taken to be the Euclidean metric $|dz|^2 = dx^2 + dy^2$ lifted to $\mathscr{M}$ by p, i.e.

$$d\tilde{s}^2 = |dp|^2 = |p'(\tilde{z})|^2(|d\tilde{x}|^2 + |d\tilde{y}|^2). \tag{4.1}$$

The intrinsic area form on $\mathscr{M}$ is similarly the pull-back of $dm = dx \wedge dy$ to $\mathscr{M}$, namely

$$d\tilde{m} = \frac{1}{2i}d\bar{p} \wedge dp = |p'(\tilde{z})|^2 d\tilde{x} \wedge d\tilde{y}. \tag{4.2}$$

In terms of the Hermitian bilinear form $d\bar{p} \otimes dp$ one can write $d\tilde{s}^2 = \text{Re}\, d\bar{p} \otimes dp$, $d\tilde{m} = \text{Im}\, d\bar{p} \otimes dp$. Indeed, identifying p with $z = x + iy$ we have

$$d\bar{p} \otimes dp = (dx - idy) \otimes (dx + idy)$$
$$= dx \otimes dx + dy \otimes dy + i(dx \otimes dy - dy \otimes dx)$$
$$= |dz|^2 + i(dx \wedge dy),$$

as viewed in the projected plane $\mathbb{C}$.

Assume now that $0 \in p(\mathcal{M})$ and let $\tilde{0} \in \mathcal{M}$ be a point such that $p(\tilde{0}) = 0$. Then we may consider Hele-Shaw evolution on $\mathcal{M}$ with injection (or suction) at $\tilde{0}$. In case of a simply connected evolution $\tilde{\Omega}(t)$, let

$$\tilde{f}(\cdot, t) : \mathbb{D} \to \tilde{\Omega}(t) \subset \mathcal{M}$$

be conformal maps with $\tilde{f}(0, t) = \tilde{0}$ and $f'(0, t) > 0$, where $f = p \circ \tilde{f}$ is the projection of $\tilde{f}$ to $\mathbb{C}$,

$$f(\zeta, t) = p(\tilde{f}(\zeta, t)).$$

The latter relationship gives

$$\frac{\dot{f}(\zeta, t)}{\zeta f'(\zeta, t)} = \frac{\dot{\tilde{f}}(\zeta, t)}{\zeta \tilde{f}'(\zeta, t)}, \tag{4.3}$$

which expresses invariance of the Poisson integral (2.11) under changes of coordinates. In particular it follows that the evolution of $\tilde{f}$ is described by

$$\dot{\tilde{f}}(\zeta, t) = \zeta \tilde{f}'(\zeta, t) P_g(\zeta, t), \tag{4.4}$$

where $P_g(\zeta, t) = P(\zeta, t)$ is the Poisson integral (2.11) defined, not in terms of $\tilde{f}'$ but in terms of $g = f'$. The relationship between f' and $\tilde{f}'$ is

$$f'(\zeta, t) = p'(\tilde{f}(\zeta, t))\tilde{f}'(\zeta, t). \tag{4.5}$$

If p is thought of as just a local identity map (away from branch points) then f' and $\tilde{f}'$ are the same.

It should be noted that $\tilde{f}(\cdot, t)$ by definition always is univalent in $\mathbb{D}$, in particular $\tilde{f}' \neq 0$ in $\mathbb{D}$. If $f' = 0$ at some point in $\mathbb{D}$, then it is the factor $p'(\tilde{f}(\zeta, t))$ in (4.5) that vanishes there. When formulated as a Polubarinova-Galin equation the evolution of $\tilde{f}$ is given by

$$\text{Re}\,[\dot{\tilde{f}}(\zeta, t)\overline{\zeta \tilde{f}'(\zeta, t)}] = \frac{q(t)}{|p'(\tilde{f}(\zeta, t))|^2} \quad (\zeta \in \partial\mathbb{D}). \tag{4.6}$$

This equation is an immediate consequence of (4.4), (4.5) and (2.11). It is actually the general form of the Polubarinova-Galin equation on a manifold with Riemannian metric given as in (4.1), even when the integral of p' is not interpreted as a covering map.

In general, $\dot{\tilde{f}}$ and $\zeta \tilde{f}'$ should be interpreted as vectors in the tangent space of $\mathcal{M}$ at $\tilde{z} = \tilde{f}(\zeta, t)$. More precisely, $\dot{\tilde{f}}$ is the speed of $\tilde{z} \in \partial\tilde{\Omega}$ when $\zeta \in \partial\mathbb{D}$ is kept fixed and $\zeta \tilde{f}'$ is a vector pointing in the outward normal direction of $\partial\tilde{\Omega}$. Equation (4.6) expresses that

$$< \dot{\tilde{f}}, \zeta \tilde{f}' >_{\mathcal{M}} = q \quad \text{on } \partial\mathbb{D}, \tag{4.7}$$

where $< \cdot, \cdot >_{\mathcal{M}}$ denotes the (real) inner product on the tangent space of $\mathcal{M}$. Alternatively, expressed in terms of the form $d\tilde{m} = \frac{1}{2i} d\bar{p} \wedge dp$, (4.6) says that

$$d\tilde{m}(\dot{\tilde{f}}, i\zeta \tilde{f}') = q \quad \text{on } \partial\mathbb{D}. \tag{4.8}$$

The antisymmetry implicit in the left member here becomes more explicit when writing the equation by means of a Poisson bracket, namely as in (2.7).

Let $G_{\tilde{\Omega}}(\tilde{z}, \tilde{0})$ be the **Green function** of $\tilde{\Omega}$, vanishing on the boundary and with behavior

$$G_{\tilde{\Omega}}(\tilde{z}, \tilde{0}) = -\log|\tilde{z}| + \text{harmonic}$$

at $\tilde{z} = \tilde{0}$. Then

$$G_{\tilde{\Omega}}(\tilde{z}, \tilde{0}) = -\log|\zeta|,$$

where $\tilde{z} = \tilde{f}(\zeta, t)$. Since $2\frac{\partial}{\partial z} \log|\zeta| = \frac{\partial}{\partial z} \log\zeta = \frac{1}{\zeta \tilde{f}'(\zeta,t)}$, this shows that (4.7), when written on the form

$$< \dot{\tilde{f}}, \frac{\zeta \tilde{f}'}{|\zeta \tilde{f}'|} >_{\mathcal{M}} = \frac{q}{|\zeta \tilde{f}'|} \quad \text{on } \partial\mathbb{D},$$

expresses that

$$\dot{\tilde{f}}_{\text{normal}} = q|\nabla G_{\tilde{\Omega}}(\tilde{z}, \tilde{0})|, \tag{4.9}$$

i.e. that the boundary moves in the outward normal direction with speed proportional to the gradient of the Green function. This is the classical description of Laplacian growth.

When $f(\cdot, t)$ solves the Löwner-Kufarev equation it is a subordination chain by Theorem 2.1 and hence it can be lifted to a Riemann surface $\mathcal{M}$ by Lemma 2.1.

Most of the previous formulas have simple formulations on $\mathcal{M}$, for example (2.2) generalizes to

$$\frac{d}{dt} \int_{\tilde{\Omega}(t)} h \, d\tilde{m} = 2\pi q(t) h(\tilde{0}), \tag{4.10}$$

for h harmonic in a neighborhood of $\tilde{\Omega}(t)$, and where $\tilde{\Omega}(t) = \tilde{f}(\mathbb{D}, t)$, $f = p \circ \tilde{f}$: $\mathbb{D} \to \mathcal{M} \to \mathbb{C}$. For subharmonic h we have inequality $\geq$. On integrating (4.10) with respect to t we arrive at the natural weak formulation of Laplacian growth on Riemannian manifolds, and at the same time a **weak formulation of the Löwner-Kufarev equation** (2.10), (2.11) (to be compared with the corresponding concept (3.4) for the Polubarinova-Galin equation):

Definition 4.1 A family of open sets $\{\tilde{\Omega}(t) \subset \mathcal{M} : t \in I\}$ with compact closure in $\mathcal{M}$ is a **weak solution** of the Laplacian growth (or Hele-Shaw) problem on $\mathcal{M}$ with a source at $\tilde{0} \in \mathcal{M}$ if, for any $s, t \in I$ with $s < t$, $\tilde{\Omega}(s) \subset \tilde{\Omega}(t)$ and

$$\int_{\tilde{\Omega}(t)} h \, d\tilde{m} - \int_{\tilde{\Omega}(s)} h \, d\tilde{m} \geq 2\pi (Q(t) - Q(s)) h(\tilde{0}) \tag{4.11}$$

holds for every $h \in SL^1(\tilde{\Omega}(t), \tilde{m})$ (see (3.6) for the notation). Formulated in terms of partial balayage (4.11) becomes

$$\text{Bal}\,(2\pi(Q(t) - Q(s))\delta_{\tilde{0}} + \chi_{\tilde{\Omega}(s)}\tilde{m}, \tilde{m}) = \chi_{\tilde{\Omega}(t)}\tilde{m} \quad (s < t). \tag{4.12}$$

See [40] for details on partial balayage on Riemannian manifolds.

4.2 Examples

Example 4.1 The purpose of this example is to prepare for the way the definition of a weak solution is to be used in the proof of Theorem 5.1. For later clarity we spell out everything quite much in detail.

Choose $s = 0$ in Definition 4.1 and assume that the domain $\tilde{\Omega}(0) = \tilde{f}(\mathbb{D}, 0) = \tilde{f}(\mathbb{D})$ at time $s = 0$ is obtained by uniformization of some fixed $f \in \mathcal{O}_{\text{norm}}(\overline{\mathbb{D}})$, as in Sect. 2.2. Thus $f = p \circ \tilde{f}$, where p is the covering map $p : \tilde{\Omega}(0) \to \mathbb{C}$ which, in terms of the trivial decomposition

$$\mathbb{D} \xrightarrow{\text{id}} \mathbb{D} \xrightarrow{f} \mathbb{C},$$

is obtained by interpreting the second $\mathbb{D}$ as an abstract Riemann surface, identified as $\tilde{\Omega}(0)$. The names of the mappings are then shifted to

$$\mathbb{D} \xrightarrow{\tilde{f}} \tilde{\Omega}(0) \xrightarrow{p} \mathbb{C}.$$

By assumption, f is analytic in some larger disk, say in $\mathbb{D}(0, \rho)$, $\rho > 1$. Thus the two diagrams extend to

$$\mathbb{D}(0, \rho) \xrightarrow{\text{id}} \mathbb{D}(0, \rho) \xrightarrow{f} \mathbb{C}, \tag{4.13}$$

$$\mathbb{D}(0, \rho) \xrightarrow{\tilde{f}} \mathcal{M} \xrightarrow{p} \mathbb{C}, \tag{4.14}$$

respectively, which defines the Riemann surface $\mathcal{M}$ as being the conformal image of $\mathbb{D}(0, \rho)$ under $\tilde{f}$. In particular, $\tilde{\Omega}(0) \subset \mathcal{M}$, and for small enough $t > 0$ the weak solution with initial domain $\tilde{\Omega}(0)$ will stay compactly in $\mathcal{M}$. The defining property of the solution domain $\tilde{\Omega}(t) \supset \tilde{\Omega}(0)$ at time $t > 0$ is, when formulated in terms of the abstract Riemann surface notations of (4.14),

$$\int_{\tilde{\Omega}(t)} \tilde{h} d\tilde{m} - \int_{\tilde{\Omega}(0)} \tilde{h} d\tilde{m} \geq 2\pi Q(t) \tilde{h}(\tilde{0}). \tag{4.15}$$

This is to hold for all integrable (with respect to $\tilde{m}$) subharmonic functions $\tilde{h}$ in $\tilde{\Omega}(t)$. When the same property is formulated by identifying $\mathcal{M}$ with $\mathbb{D}(0, \rho)$ as in (4.13) it becomes

$$\int_{D(t)} h|g|^2 dm - \int_{\mathbb{D}} h|g|^2 dm \geq 2\pi Q(t) h(0), \tag{4.16}$$

to hold for all subharmonic h in $D(t)$, integrable with respect to the measure $|g|^2 dm$. Here $D(t) = \tilde{f}^{-1}(\tilde{\Omega}(t)) \subset \mathbb{D}(0, \rho)$, $g = f'$, and $h = \tilde{h} \circ \tilde{f}$, which is subharmonic if and only if $\tilde{h}$ is. Note that (by definition, (4.1)) $d\tilde{m} = p^*(dm)$ in the picture (4.14), which becomes $|g|^2 dm$ in the picture (4.13).

The domains $\tilde{\Omega}(t)$ and $D(t)$ are not necessarily simply connected when $t > 0$, as they are defined only in terms of a weak solution. Eventually, however, we want to assert that they are simply connected if $t > 0$ is small enough.

The only thing which can make a weak solution break down is that it runs out of the manifold, $\mathcal{M}$. Then the natural thing to do is to try to extend $\mathcal{M}$ to a larger manifold. Weak solutions are unique (up to null-sets), but they of course depend on the choice of $\mathcal{M}$ and p. If we take $\mathcal{M}$ to be, for example, a disk $\mathbb{D}(0, a) \subset \mathbb{C}$ then, assuming $Q(t) \to \infty$ as $t \to \infty$, any Hele-Shaw evolution will eventually run out of $\mathcal{M}$. The following example shows that there are always many different ways of enlarging $\mathcal{M}$, which then give rise to different evolutions.

Example 4.2 Choose a point $a > 0$ on the positive real axis, to be used as a stopping point and also as a branch point. Let $\mathcal{M} = \mathbb{D}(0, a)$ be the disk reaching out to a, and consider it as a Riemann surface with trivial projection map $p(z) = z$ to $\mathbb{C}$. Then starting from empty space, a Hele-Shaw flow evolution on $\mathcal{M}$ with injection at the origin gives a family of growing disks, say $\Omega(t) = \mathbb{D}(0, at)$ on the time interval $0 < t < 1$, as a weak (and strong) solution for the source strength $q(t) = a^2 t$

(so that $2\pi Q(t) = m(\mathbb{D}(0, at)))$. At time $t = 1$ it runs out of $\mathcal{M}$, but it can be continued without any changes on the trivially extended Riemann surface $\mathcal{M}_1 = \mathbb{C}$, for $0 < t < \infty$.

However, it can also be continued in many other ways. Let for example $\mathbb{C}_1$, $\mathbb{C}_2$ be two copies of $\mathbb{C}$ and consider

$$\mathcal{M}_2 = (\mathbb{C}_1 \setminus \{a\}) \cup (\mathbb{C}_2 \setminus \{a\}) \cup \{a\}$$

as a covering surface of $\mathbb{C}$ with a branch point at $z = a$. The covering map $p :$ $\mathcal{M}_2 \to \mathbb{C}$ identifies any point on $\mathbb{C}_j$ ($j = 1, 2$) with the corresponding point on $\mathbb{C}$. This is also true at $z = a$, but a more accurate description there has to be given in terms of a local coordinate. We may for example choose a local coordinate $\tilde{z}$ on $\mathcal{M}_2$ so that $\tilde{z} = 0$ corresponds to $z = a$ and, more precisely, so that

$$p_2(\tilde{z}) = \tilde{z}^2 + a.$$

Thus, with $z = p_2(\tilde{z})$, $\tilde{z} = \sqrt{z - a}$. This coordinate $\tilde{z}$ is actually a global coordinate on $\mathcal{M}_2$ and it makes $\mathcal{M}_2$ appear as the classical Riemann surface of the multivalued function $\sqrt{z - a}$ (Figs. 4.1, 4.2).

In terms of the above coordinate, the area form of $\mathcal{M}_2$ is

$$d\tilde{m}_2 = \frac{1}{2\mathrm{i}} d\bar{p}_2 \wedge dp_2 = 4|\tilde{z}|^2 d\tilde{x} d\tilde{y}.$$

The source point is to be one of the two points $\pm\sqrt{-a}$ on $\mathcal{M}_2$ above $0 \in \mathbb{C}$, let it be $\tilde{0} = \mathrm{i}\sqrt{a}$, $\sqrt{a}$ denoting the positive root. Now the definition of a weak solution on $\mathcal{M}_2$ becomes, explicitly,

$$4 \int_{\tilde{\Omega}(t)} h(\tilde{z})|\tilde{z}|^2 d\tilde{x} d\tilde{y} - 4 \int_{\tilde{\Omega}(s)} h(\tilde{z})|\tilde{z}|^2 d\tilde{x} d\tilde{y} \geq 2\pi(t - s)h(\mathrm{i}\sqrt{a}),$$

to hold for all integrable (with respect to $\tilde{m}_2$) subharmonic functions h in $\tilde{\Omega}(t)$. Expressed in the coordinate $\tilde{z}$ it is thus a weighted Hele-Shaw flow, as discussed in for example [53]. It exists for all $0 < t < \infty$, but it is certainly different from the solution $\Omega(t)$ on $\mathcal{M}_1 = \mathbb{C}$. For $t > 1$, and when viewed on $\mathcal{M}_2$, part of $\tilde{\Omega}(t)$ continues on the original ('lower') sheet, say $\mathbb{C}_1$, while part leaks to the 'upper' sheet $\mathbb{C}_2$. Compare Fig. 4.3. Hedenmalm and Shimorin [53] use the terminology 'wrapped Hele-shaw flow' when the solution goes up on a Riemann covering surface, at least in the case when there are no branch points.

Example 4.3 This example can be viewed as a continuation of Example 4.2 (and also of Example 2.1), but from a different point of view. It is based on an example

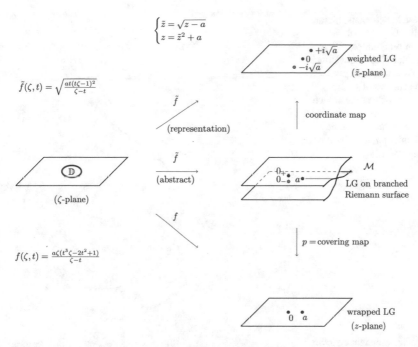

$$\begin{cases} \tilde{z} = \sqrt{z - a} \\ z = \tilde{z}^2 + a \end{cases}$$

weighted LG
($\tilde{z}$-plane)

$\tilde{f}(\zeta, t) = \sqrt{\frac{at(t\zeta - 1)^2}{\zeta - t}}$

coordinate map

$\tilde{f}$
(representation)

$\tilde{f}$
(abstract)

$\mathcal{M}$
LG on branched
Riemann surface

(ζ-plane)

f

$p =$ covering map

$f(\zeta, t) = \frac{a\zeta(t^3\zeta - 2t^2 + 1)}{\zeta - t}$

wrapped LG
(z-plane)

Fig. 4.1 Three ways of viewing the nonunivalent evolution in Example 4.2. The expressions for $\tilde{f}$ and f refer to the regime $t > 1$. Choice of squareroot can be made (for example) such that injection point corresponds to $\tilde{z} = -i\sqrt{a}$ and, on the abstract Riemann surface, to 0_-. The regions of fluid in the $\tilde{z}$- and z-planes are illustrated in Figs. 4.2 and 4.3, respectively

of Sakai [100], and it appears also, from a different point of view, in [48]. Let

$$f(\zeta, t) = b(t) \frac{\zeta(\zeta - 2t^{-1} + t^{-3})}{\zeta - t},$$

where $1 < t < \infty$ and $b(t) \in \mathbb{R}$ are parameters. The derivative is

$$g(\zeta, t) = b(t) \frac{(\zeta - t^{-1})(\zeta - 2t + t^{-1})}{(\zeta - t)^2}. \qquad (4.17)$$

We see that g has two zeros, $\omega_1(t) = t^{-1} \in \mathbb{D}$ and $\omega_2(t) = 2t - t^{-1} \in \mathbb{C} \setminus \overline{\mathbb{D}}$, and that $g(\zeta, t)d\zeta$, as a differential, has double poles at $\zeta_1(t) = t = \frac{1}{2}(\omega_1(t) + \omega_2(t))$ and at infinity. The data of g are special in two ways: first of all ω_1 and ζ_1 are reflections of each other with respect to the unit circle, and secondly ω_2 is chosen so that $gd\zeta$ has no residues (which is immediately clear since f has no logarithmic poles).

Since $\omega_1(t) \in \mathbb{D}$, $f(\cdot, t)$ is not locally univalent in $\mathbb{D}$, but it generates the same moments as a disk: all moments (defined by the rightmost member in (2.4)) vanish,

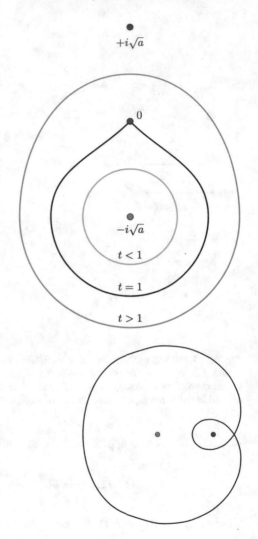

Fig. 4.2 Schematic picture, inspired by [11], of the evolution Example 4.2 as weighted Hele-Shaw flow in the $\bar{z}$-plane (see Fig. 4.1), for three choices of t. The interior angle at the branch point is 90 degrees, and the boundary moves at that particular point with infinite speed

Fig. 4.3 Schematic picture of evolution in Example 4.2 as wrapped Hele-Shaw flow in the z-plane (see Fig. 4.1) when $t > 1$. The orange dot is the point of injection (the origin) and the blue dot is the branch point a. The small area around the branch point is covered twice

except the first one which is

$$M_0(t) = b(t)^2 \frac{2t^2 - 1}{t^4}.$$

More generally, the corresponding quadrature identity is

$$\frac{1}{\pi} \int_{\mathbb{D}} h(\zeta)|g(\zeta, t)|^2 dm(\zeta) = b(t)^2 \frac{2t^2 - 1}{t^4} h(0), \qquad (4.18)$$

holding for h analytic and integrable in $\mathbb{D}$. This formula also shows that, for the special choice $b(t) = \frac{t^2}{\sqrt{2t^2-1}}$, $g(\zeta, t)$ is a contractive (inner) zero divisor in the sense of Hedenmalm [48, 49].

Despite (4.18), $f(\zeta, t)$ in general does not solve the Polubarinova-Galin equation (2.1). Only for one particular choice of $b(t)$ it does. This choice is determined by the requirement that $f(\omega_1(t), t)$ shall be time independent. Since

$$f(\omega_1(t), t) = f(t^{-1}, t) = \frac{b(t)}{t^3}$$

this condition gives

$$b(t) = at^3, \tag{4.19}$$

where a is a constant. A calculation shows that for this particular choice of $b(t)$, the Polubarinova-Galin equation indeed holds, with $q(t) = a^2 t(4t^2 - 1)$:

$$\mathrm{Re}\left[\dot{f}(\zeta, t)\overline{\zeta f'(\zeta, t)}\right] = a^2 t(4t^2 - 1) \quad \text{for } \zeta \in \partial\mathbb{D}.$$

Note that $q(t) > 0$. Also the Löwner-Kufarev equation holds, because $f(\omega_1(t), t) = a$ is fixed (cf. Theorem 2.1).

Now we shall see that, taking $a > 0$, the above solution, namely

$$f(\zeta, t) = \frac{a\zeta(t^3\zeta - 2t^2 + 1)}{\zeta - t},$$

is exactly the projection under p_2 of the evolution on $\mathcal{M}_2$ in Example 4.2. In fact, since $f(\cdot, t)$ maps the zero $\omega_1(t) \in \mathbb{D}$ of $g(\zeta, t)$ onto the fixed point a, $f(\cdot, t)$ lifts to a map $\tilde{f}(\cdot, t)$ into the surface $\mathcal{M}_2$. Inverting $p_2(\tilde{z}) = \tilde{z}^2 + a$ gives the explicit expression

$$\tilde{f}(\zeta, t) = \sqrt{f(\zeta, t) - a} = \sqrt{\frac{at(t\zeta - 1)^2}{\zeta - t}}.$$

This function, for any fixed $t > 1$, is univalent, $\tilde{f}(\cdot, t) : \mathbb{D} \to \tilde{\Omega}(t) \subset \mathcal{M}_2$, and as a function of t it represents, in the coordinate $\tilde{z}$, the Hele-Shaw evolution on $\mathcal{M}_2$. Indeed, it satisfies the Polubarinova-Galin equation on $\mathcal{M}_2$:

$$\mathrm{Re}\left[\dot{\tilde{f}}(\zeta, t)\overline{\zeta \tilde{f}'(\zeta, t)}\right] = \frac{q(t)}{4|\tilde{f}(\zeta, t)|^2} \quad \text{for } \zeta \in \partial\mathbb{D}.$$

This is an instance of (4.6), as $p_2'(\tilde{z}) = 2\tilde{z}$.

As a summary, we write up in coordinates, z and $\tilde{z}$, and $0 < t < \infty$, the complete evolution in Example 4.2, namely the growing disk which at the point a climbs up to the Riemann surface $\mathcal{M}_2$:

(i) In terms of z, solving the ordinary Löwner-Kufarev equation (2.10), (2.11),

$$f(\zeta, t) = \begin{cases} a t \zeta & (0 < t < 1), \\ \dfrac{a\zeta(t^3\zeta - 2t^2 + 1)}{\zeta - t} & (1 < t < \infty). \end{cases} \qquad (4.20)$$

Notice that each of the two expressions above is (real) analytic in t, even across the junction value $t = 1$. Thus the combined function $f(\zeta, t)$ is piecewise real analytic with respect to t.

(ii) In terms of $\tilde{z}$, for which we have (4.4) and (4.6) when $t \neq 1$, and for which the entire solution (across $t = 1$) is a weak solution on $\mathcal{M}_2$,

$$\tilde{f}(\zeta, t) = \begin{cases} \sqrt{a(t\zeta - 1)} & (0 < t < 1), \\ \sqrt{\dfrac{at(t\zeta - 1)^2}{\zeta - t}} & (1 < t < \infty). \end{cases}$$

The source strength is

$$q(t) = \begin{cases} a^2 t & (0 < t < 1), \\ a^2 t(4t^2 - 1) & (1 < t < \infty). \end{cases} \qquad (4.21)$$

Here we can see a discontinuity of $q(t)$ at $t = 1$. However this is harmless, and can be avoided by using another time parametrization. For example one could define $f(\zeta, t) = at^3\zeta$, $\tilde{f}(\zeta, t) = \sqrt{a(t^3\zeta - 1)}$ for $0 < t < 1$, which gives the same family of domains, just traversed with a different speed. This would give $q(t) = 3a^2 t^5$ for $0 < t < 1$, making $q(t)$ continuous across $t = 1$.

Example 4.4 As another description of the solutions, one can easily write up the equations for the boundary curves $f(\partial \mathbb{D}, t)$ explicitly. Indeed, since each $f(\zeta, t)$ is a rational function the boundary curves are algebraic, of the form $Q_t(z, \bar{z}) = 0$ for some polynomials $Q_t(z, w)$ which can be found via the elimination function in Sect. 3.6.

For $0 < t < 1$ it is clear that the boundary is the circle $|z|^2 = a^2 t^2$, thus defined by the equation $Q_t(z, \bar{z}) = 0$ where $Q_t(z, w) = zw - a^2 t^2$. In general, the defining polynomial $Q_t(z, w)$ appears in the numerator in the elimination function

$$\mathcal{E}_{f, f^*}(z, \bar{w}) = \frac{Q_t(z, \bar{w})}{P_t(z)\overline{P_t(w)}},$$

see (3.33). By direct calculation of the resultant involved in the elimination function one finds that, for $t > 1$,

$$Q_t(z, w) = z^2 w^2 - a\,zw(z + w) - a^2(2t^4 - t^2 - 1)\,zw$$
$$+ a^3 t^2(2t^2 - 1)(z + w) + a^4 t^4(t^4 - 3t^2 + 1),$$
$$P_t(z) = z(z - a).$$

The transition between the phases $t < 1$ and $t > 1$ shows up clearly in the elimination functions:

$$\mathscr{E}_t(z, w) = \frac{zw - a^2 t^2}{zw} \qquad \text{for } 0 < t < 1,$$

$$\mathscr{E}_t(z, w) = \frac{(t^2 - 1)^2 Q_t(z, w)}{zw(z - a)(w - a)} \qquad \text{for } t > 1.$$

When $t = 1$ the polynomial $Q_t(z, w)$ factorizes as

$$Q_t(z, w) = (zw - a^2)(z - a)(w - a),$$

hence the transition is smooth except for the innocent the factor $(t^2 - 1)^2$.

Since $Q_t(z, w)$ is a polynomial of degree one (for $t < 1$) or two (for $t > 1$) one can easily solve the equation $Q_t(z, w) = 0$ and thereby obtain the Schwarz function $w = S_t(z)$ for the boundary curve. For $t < 1$ the result is $S_t(z) = a^2 t^2/z$, while for $t > 1$ one gets the somewhat lengthy expression

$$S_t(z) = \frac{1}{2z(z - a)}\left(az^2 + a^2(2t^4 - t^2 - 1)z - a^3 t^2(2t^2 - 1) \pm \sqrt{R_t(z)}\right),$$

where

$$R_t(z) = a^2 z^4 + (-4t^2 + 2t^2 - 2)a^3 z^3 + (4t^6 + 5t^4 - 4t^2 + 1)a^4 z^2$$
$$+ (-4t^8 - 4t^6 + 2t^4 - 2t^2 + 4)a^5 z + (4t^8 - 4t^6 + t^4)a^6.$$

It is interesting to notice that $S_t(z)$ has a pole not only at $z = 0$, but also at the branchpoint $z = a$ when $t > 1$. When $t \searrow 1$ one of the two branches of the Schwarz functions becomes $S(z) = a^2/z$, i.e. the same as when appraoching $t = 1$ from $t < 1$, while the other branch becomes the constant $S(z) = a$.

4.3   The Riemann Surface Solution Pulled Back to the Unit Disk

For a function $h(\zeta, t)$ which is holomorphic in ζ the requirement (3.1), with $\Psi = h$, reduces to the simpler statement

$$\dot{h}(\zeta, t) f'(\zeta, t) = \dot{f}(\zeta, t) h'(\zeta, t). \tag{4.22}$$

This can be viewed as the vanishing of a functional determinant and can alternatively be written as

$$\frac{\dot{h}(\zeta, t)}{\zeta h'(\zeta, t)} = \frac{\dot{f}(\zeta, t)}{\zeta f'(\zeta, t)}, \tag{4.23}$$

where (on dividing by ζ) we also have used that $f(0, t) = 0$ for all t. When f solves the Löwner-Kufarev equation (2.10) the right member is holomorphic in $\mathbb{D}$ and equals $P(\zeta, t)$. Then (4.23) means that h solves the same Löwner-Kufarev equation as f. This can be interpreted as saying that 'h flows with f', and it also follows that the $h(\zeta, t)$ are subordinated by the same functions as $f(\zeta, t)$:

$$h(\varphi(\zeta, s, t), t) = h(\zeta, s) \quad (s \le t). \tag{4.24}$$

Here $\varphi(\zeta, s, t)$ are the subordination functions in (2.23). Note that (4.24), or (4.22), implies that $h(0, t) = h(0, s)$ (or $\dot{h}(0, t) = 0$).

We can now assert

Proposition 4.1 *Let* $t \mapsto f(\cdot, t) \in \mathscr{O}_{norm}(\mathbb{D})$ *be a smooth evolution on some time interval and assume that* $f' \ne 0$ *on* $\partial \mathbb{D}$ *on this time interval. Then* $f(\cdot, t)$ *solves the Polubarinova-Galin equation (2.1) if and only if*

$$\frac{d}{dt} \int_{\mathbb{D}} h(\zeta, t) |f'(\zeta, t)|^2 \, dm(\zeta) = 2\pi q(t) h(0, t). \tag{4.25}$$

for every function $h(\cdot, t) \in \mathscr{O}(\overline{\mathbb{D}})$ *which satisfies (4.22) (equivalently, (3.1) or (4.24)), and it solves the Löwner-Kufarev equation (2.10) if and only if moreover (2.24) holds (equivalently,* $f(\cdot, t)$ *is a subordination chain).*

Proof The proposition follows immediately from Lemma 3.1 and Theorem 2.1 since Re $h|_{\partial \mathbb{D}}$ and Im $h|_{\partial \mathbb{D}}$ range over a dense set of functions in (4.25).   □

Also the Riemann surface weak formulation (4.11) can, in case relevant domains are simply connected, be pulled back to the unit disk in various ways. In Example 4.1 this was done by pulling the initial domain back to $\mathbb{D}$, which works well for discussing solutions on a short time interval $0 \le t < \varepsilon$. However, to discuss global solutions it is better to fix a final time $t = T$ under consideration, and then pull back the domain $\tilde{\Omega}(T) \subset \mathcal{M}$ at that time to $\mathbb{D}$, assuming that $\tilde{\Omega}(T)$ is simply connected. Then all previous domains become subdomains of $\mathbb{D}$.

Thus fixing T and identifying $\tilde{\Omega}(T)$ with $\mathbb{D}$ via $\tilde{f}(\cdot, T)$, Eq. (4.11) becomes, for $s < t \leq T$ and on setting $g = f'$ as usual,

$$\int_{D(t,T)} h(z)|g(z,T)|^2 \, dm(z) - \int_{D(s,T)} h(z)|g(z,T)|^2 \, dm(z)$$

$$\geq 2\pi (Q(t) - Q(s))h(0), \tag{4.26}$$

to hold for $h \in SL^1(D(t,T), m)$. Here the domains $D(s,T) = \tilde{f}^{-1}(\tilde{\Omega}(s), T)$, $D(t,T) = \tilde{f}^{-1}(\tilde{\Omega}(t), T)$, satisfying $D(s,T) \subset D(t,T) \subset \mathbb{D}$, need not be simply connected. Choosing $t = T = 0$ with $s < 0$ gives

$$\int_{\mathbb{D}} h(z)|g(z,0)|^2 \, dm(z) - \int_{D(s)} h(z)|g(z,0)|^2 \, dm(z) \geq -2\pi Q(s)h(0), \tag{4.27}$$

where $D(s) = D(s,0) \subset \mathbb{D}$ and $Q(s) < 0$. This is a counterpart of (4.16) for negative times. It also connects to the theory of finite contractive zero divisors: starting, as in Example 4.2, a Hele-Shaw evolution on a Riemann surface $\mathcal{M}$ from empty space, we have $D(s) = \emptyset$ at the initial time $s < 0$, and then (4.27) can be identified with the definition of an inner divisor (namely $g(z,0)$ in the above equation), as in [48, 49].

In terms of partial balayage, (4.26) with $s = 0$ takes the form

$$\mathrm{Bal}\,(2\pi Q(t)\delta_0 + |g(\cdot,T)|^2 \chi_{D(0,T)}, |g(\cdot,T)|^2 \chi_{\mathbb{D}}) = |g(\cdot,T)|^2 \chi_{D(t,T)}.$$

The weak solution can be coupled to the Löwner-Kufarev equation only if the domains $\tilde{\Omega}(t)$, or $D(t,T)$, are simply connected. When this is the case we have $D(t,T) = \varphi(\mathbb{D}, t, T)$, where $\varphi(\zeta, s, t)$ are the subordination functions associated to the conformal maps $f(\cdot, t) : \mathbb{D} \to \tilde{\Omega}(t)$. In such a case, and returning to (4.26), choosing $T = t$ there and making the variable transformation $z = \varphi(\zeta, s, t)$ in the last integral, one gets

$$\int_{\mathbb{D}} h(z)|g(z,t)|^2 \, dm(z) - \int_{\mathbb{D}} h(\varphi(\zeta,s,t))|g(\zeta,s)|^2 \, dm(\zeta)$$

$$\geq 2\pi (Q(t) - Q(s))h(0), \tag{4.28}$$

to hold for h subharmonic and integrable in $\mathbb{D}$. For harmonic h we have equalities in the above inequalities because both of $\pm h$ are then subharmonic. The relation (4.28) can also be obtained directly by integrating (4.25) and using (4.24). Note that for time dependent test functions, $h(z, t)$, which satisfy (4.24), the relation (4.28) takes the simpler form

$$\int_{\mathbb{D}} h(z,t)|g(z,t)|^2 \, dm(z) - \int_{\mathbb{D}} h(z,s)|g(z,s)|^2 \, dm(z) \geq$$

$$\geq 2\pi(Q(t) - Q(s))h(0,t). \qquad (4.29)$$

As for the right member, $h(0,t)$ is actually independent of t by (4.24).

We summarize:

Proposition 4.2 *A family* $\{f(\cdot,t) \in \mathscr{O}_{norm}(\overline{\mathbb{D}}) : 0 \leq t \leq T\}$ *represents a weak solution as in Definition 4.1 (with* $I = [0,T]$*) if an only if it is a subordination chain as in Definition 2.2 and (4.28) holds for* $0 \leq s < t \leq T$.

4.4 Compatibility Between Balayage and Covering Maps

The family of branched covering surfaces over $\mathbb{C}$ form a partially ordered set in a natural way. Within in each of the surfaces one can perform partial balayage, sweeping to the area form lifted from $\mathbb{C}$. Thus we have two kinds of projection maps, reducing refined objects to cruder objects containing less information:

(i) The first is the balayage operator taking, for example, an initial domain $\Omega(0)$ to the domain at a later time $\Omega(t)$ by sweeping out the accumulated source:

$$\mathrm{Bal}\,(2\pi\,Q(t)\delta_{\tilde{0}} + \chi_{\Omega(s)}\tilde{m}, \tilde{m}) = \chi_{\Omega(t)}\tilde{m}.$$

This map is really an orthogonal projection in a Hilbert space (e.g. the Sobolev space $H_0^1(\mathscr{M}) = W_0^{1,2}(\mathscr{M})$ if the Dirac measures are suitably smoothed out). It is a 'horizontal' projection, within each covering surface.

(ii) The second is the branched covering map between two Riemann surfaces, by which a measure on the higher surface can be pushed down to a measure on the lower surface. One may think of this as a 'vertical' projection.

The aim of the present section is to show that these two projections commute in an appropriate sense. Let $p : \mathscr{M} \to \mathbb{C}$ be a branched covering map, i.e. p is a nonconstant analytic function. By p_* we denoted the push-forward map, which can be applied to measures on $\mathscr{M}$, to (parametrized) chains for integration (simply by composition), etc. Similarly, p^* denotes the pull-back map, which can be applied to functions and differential forms on $\mathbb{C}$. If for example $\tilde{\Omega}$ is a domain in $\mathscr{M}$, thought of as the oriented 2-chain parametrized by some $\tilde{f} : \mathbb{D} \to \mathscr{M}$ (where $\tilde{\Omega} = \tilde{f}(\mathbb{D})$), then $p_*\tilde{\Omega}$ is the 2-chain parametrized by $f = p \circ \tilde{f} : \mathbb{D} \to \mathbb{C}$, which can be thought of as $\Omega = f(\mathbb{D})$ with appropriate multiplicities. In other words, p_* takes the measure $\chi_{\tilde{\Omega}}\tilde{m}$, on $\mathscr{M}$, where $d\tilde{m} = p^*dm = d(p \circ x) \wedge d(p \circ y)$, to $\nu_f m$ on $\mathbb{C}$, ν_f being the counting function, Definition 2.1.

Note that p_* and p^* are linear maps on suitable vector spaces and that they, in some formal sense, are each others adjoints. For example, for measures μ with compact support on $\mathscr{\tilde{M}}$ and continuous functions φ on $\mathbb{C}$ we have

$$\int_{\mathbb{C}} \varphi \, d(p_* \mu) = \int_{\mathscr{\tilde{M}}} (\varphi \circ p) \, d\mu = \int_{\mathscr{\tilde{M}}} (p^* \varphi) \, d\mu.$$

The first identity here can be used as a definition of p_* when it acts on measures, and $p^*(\varphi)$ is simply defined as $\varphi \circ p$. See further Sections 2.5.19 and 4.1.7 in [21].

In order to be able to use systematic notations we now denote the complex plane by $\mathscr{M}$, and we call the covering surface $\mathscr{\tilde{M}}$. This makes the proposition below look like a quite general result (which it in fact is, but we shall only prove it under the stated assumptions). Recall that a **proper** map between topological spaces is a map p such that the inverse image $p^{-1}(K)$ of any compact set K is itself compact. That p is proper means essentially that p maps the boundary to the boundary.

Proposition 4.3 With $p : \mathscr{\tilde{M}} \to \mathscr{M}$ a nonconstant proper analytic map between two Riemann surfaces, where $\mathscr{M} = \mathbb{C}$, let $\tilde{\mu}$ be a measure with compact support in $\mathscr{\tilde{M}}$, λ a measure on $\mathscr{M}$, absolutely continuous with respect to m and satisfying (3.7), and let $\tilde{\lambda}$ be a measure on $\mathscr{\tilde{M}}$ satisfying $\tilde{\lambda} \geq p^* \lambda$. Then

$$Bal\,(p_* Bal\,(\tilde{\mu}, \tilde{\lambda}), \lambda) = Bal\,(p_* \tilde{\mu}, \lambda).$$

Proof Since $\mathscr{M} = \mathbb{C}$ and p is proper, $\mathscr{\tilde{M}}$ will be large enough for $Bal\,(\tilde{\mu}, \tilde{\lambda})$ to exist and have compact support in $\mathscr{\tilde{M}}$. Set then

$$\tilde{v} = Bal\,(\tilde{\mu}, \tilde{\lambda}),$$
$$v' = Bal\,(p_* \tilde{v}, \lambda),$$
$$\mu = p_* \tilde{\mu},$$
$$v = Bal\,(\mu, \lambda),$$

and we shall show that $v' = v$.

By the general structure of partial balayage (3.10) we have

$$\tilde{v} = \tilde{\lambda} \chi_{\tilde{\Omega}} + \tilde{\mu} \chi_{\mathscr{\tilde{M}} \setminus \tilde{\Omega}}, \tag{4.30}$$

where $\tilde{\Omega} \subset \mathscr{\tilde{M}}$ is the maximal open set in which $\tilde{v} = \tilde{\lambda}$. Recall from (3.12), (3.13) that this $\tilde{\Omega}$ can also be characterized by

$$\begin{cases} \tilde{\mu} < \tilde{\lambda} \text{ on } \mathscr{\tilde{M}} \setminus \tilde{\Omega}, \\ \int_{\tilde{\Omega}} \psi \, d\tilde{\mu} \leq \int_{\tilde{\Omega}} \psi \, d\tilde{\lambda} \text{ for all } \psi \in SL^1(\tilde{\Omega}, \tilde{\lambda}). \end{cases} \tag{4.31}$$

Here $SL^1(\tilde{\Omega}, \tilde{\lambda})$ may be replaced by a smaller test class, as discussed after (3.12), (3.13), to avoid some possible integrability problems below.

Similarly to the above we have

$$\nu' = \lambda \chi_{\Omega'} + (p_* \tilde{\nu}) \, \chi_{\mathcal{M} \setminus \Omega'}, \tag{4.32}$$

where $\Omega' \subset \mathcal{M}$ is characterized by

$$\begin{cases} p_* \tilde{\nu} < \lambda \text{ on } \mathcal{M} \setminus \Omega', \\ \int_{\Omega'} \varphi \, d(p_* \tilde{\nu}) \le \int_{\Omega'} \varphi \, d\lambda \quad (\varphi \in SL^1(\Omega', \lambda)) \end{cases} \tag{4.33}$$

and

$$\nu = \lambda \chi_\Omega + \mu \, \chi_{\mathcal{M} \setminus \Omega}, \tag{4.34}$$

with $\Omega \subset \mathcal{M}$ characterized by

$$\begin{cases} \mu < \lambda \text{ on } \mathcal{M} \setminus \Omega, \\ \int_\Omega \varphi \, d\mu \le \int_\Omega \varphi \, d\lambda \quad (\varphi \in SL^1(\Omega, \lambda)). \end{cases}$$

Since p_* is a linear operator (4.30) gives

$$p_* \tilde{\nu} = p_*(\tilde{\lambda} \chi_{\tilde{\Omega}}) + p_*(\tilde{\mu} \chi_{\tilde{\mathcal{M}} \setminus \tilde{\Omega}}). \tag{4.35}$$

By the assumption $\tilde{\lambda} \ge p^*(\lambda)$ we have $p_*(\tilde{\lambda} \chi_{\tilde{\Omega}}) \ge \lambda \chi_{p(\tilde{\Omega})}$. Thus (4.35) shows that $p_* \tilde{\nu} \ge \lambda$ in $p(\tilde{\Omega})$. It follows that $\nu' \ge \lambda$ in $p(\tilde{\Omega})$, hence

$$p(\tilde{\Omega}) \subset \Omega'. \tag{4.36}$$

By definition of $p_* \tilde{\nu}$, the second part of (4.33) spells out to

$$\int_{p^{-1}(\Omega')} (\varphi \circ p) \, d\tilde{\nu} \le \int_{\Omega'} \varphi \, d\lambda \quad (\varphi \in SL^1(\Omega', \lambda)),$$

which in view of (4.30) gives that

$$\int_{p^{-1}(\Omega') \cap \tilde{\Omega}} (\varphi \circ p) \, d\tilde{\lambda} + \int_{p^{-1}(\Omega') \setminus \tilde{\Omega}} (\varphi \circ p) \, d\tilde{\mu} \le \int_{\Omega'} \varphi \, d\lambda \quad (\varphi \in SL^1(\Omega', \lambda)).$$

Next we take $\psi = p^* \varphi = \varphi \circ p$ in (4.31). This gives

$$\int_{\tilde{\Omega}} (\varphi \circ p) \, d\tilde{\mu} \le \int_{\tilde{\Omega}} (\varphi \circ p) \, d\tilde{\lambda} \quad (\varphi \in SL^1(\Omega', \lambda)).$$

Combining with the previous inequality and using that $p^{-1}(\Omega') \supset \tilde{\Omega}$ by (4.36), gives, for $\varphi \in SL^1(\Omega', \lambda)$,

$$\int_{\Omega'} \varphi \, d\mu = \int_{p^{-1}(\Omega')} (\varphi \circ p) \, d\tilde{\mu} = \int_{\tilde{\Omega}} (\varphi \circ p) \, d\tilde{\mu} + \int_{p^{-1}(\Omega') \setminus \tilde{\Omega}} (\varphi \circ p) \, d\tilde{\mu}$$

$$\leq \int_{\tilde{\Omega}} (\varphi \circ p) \, d\tilde{\lambda} + \int_{p^{-1}(\Omega') \setminus \tilde{\Omega}} (\varphi \circ p) \, d\tilde{\mu} \leq \int_{\Omega'} \varphi \, d\lambda \quad (\varphi \in SL^1(\Omega', \lambda)).$$

In summary,

$$\int_{\Omega'} \varphi \, d\mu \leq \int_{\Omega'} \varphi \, d\lambda \quad (\varphi \in SL^1(\Omega', \lambda)). \tag{4.37}$$

We also have, by (4.30), (4.33) and, respectively, (4.36),

$$p_*(\tilde{\mu} \chi_{\tilde{\mathcal{M}} \setminus \tilde{\Omega}}) \leq p_* \tilde{\nu} < \lambda \quad \text{on } \mathcal{M} \setminus \Omega',$$

$$p_*(\tilde{\mu} \chi_{\tilde{\Omega}}) = 0 \quad \text{in } \mathcal{M} \setminus \Omega'.$$

Therefore $\mu = p_* \tilde{\mu} < \lambda$ on $\mathcal{M} \setminus \Omega'$. In combination with (4.37) this gives

$$\nu = \lambda \chi_{\Omega'} + \mu \chi_{\mathcal{M} \setminus \Omega'}.$$

Now (4.32), (4.35), (4.36) finally give

$$\nu' = \lambda \chi_{\Omega'} + (p_* \tilde{\nu}) \, \chi_{\mathcal{M} \setminus \Omega'} = \lambda \chi_{\Omega'} + (p_*(\tilde{\lambda} \chi_{\tilde{\Omega}}) + p_*(\tilde{\mu} \chi_{\tilde{\mathcal{M}} \setminus \tilde{\Omega}})) \chi_{\mathcal{M} \setminus \Omega'}$$

$$= \lambda \chi_{\Omega'} + (p_* \tilde{\mu}) \, \chi_{\mathcal{M} \setminus \Omega'} = \lambda \chi_{\Omega'} + \mu \, \chi_{\mathcal{M} \setminus \Omega'} = \nu,$$

as desired.

$\square$

Chapter 5
Global Simply Connected Weak Solutions

Abstract The major assertion (Theorem 5.1) in this chapter is that starting Laplacian growth from any simply connected planar domain with smooth boundary, the solution, in a weak form, can be continued forever as an evolution of simply connected domains. The price for keeping the domains simply connected is that the solution must be allowed to go up on a branched Riemann surface above the complex plane, a Riemann surface which is not known in advance and has to be created along with the solution. The proof of Theorem 5.1 depends on several technical lemmas, one of which (Lemma/Conjecture 5.3) is stated as a conjecture. There is no real doubt concerning the validity of that lemma/conjecture, but at present a rigorous proof is missing.

5.1 Statement of Result, and Two Lemmas

As already mentioned, given a covering map $p : \mathscr{M} \to \mathbb{C}$ as in Chap. 4 and any initial domain $\tilde{\Omega}(0) \subset \mathscr{M}$ with $\tilde{0} \in \Omega(0)$, a unique global weak solution $\{\tilde{\Omega}(t) : 0 \leq t < \infty\}$, in the sense of Definition 4.1, exists if just $\mathscr{M}$ is large enough. And if $\mathscr{M}$ is not large enough from outset it may always be extended, in many ways (compare Example 4.2), to allow for such a global weak solution. However, even if the initial domain $\tilde{\Omega}(0)$ is simply connected the weak solution will in general not remain simply connected all the time.

Now our statement, Theorem 5.1, asserts that if $\tilde{\Omega}(0)$ is simply connected and has analytic boundary, then it is indeed always possible to choose $\mathscr{M} \supset \tilde{\Omega}(0)$ so that the solution $\tilde{\Omega}(t)$ in $\mathscr{M}$ remains simply connected all the time. Without referring to any Riemann surface the assertion may be formulated simply as saying that there exists a global weak solution of the Löwner-Kufarev equation, for any given $f(\cdot, 0) \in \mathscr{O}_{\mathrm{norm}}(\mathbb{D})$. The solution cannot not always be smooth in t, because if zeros of $g = f'$ reach the unit circle then it is in most cases necessary to change the structure of g in order to make the solution go on. The Riemann surfaces involved are needed mainly to make the appropriate notion of a weak solution precise (Definition 4.1).

The difficulty in constructing $\mathscr{M}$ lies in the fact that it cannot be constructed right away, but has to be created along with the solution. It has to be updated every

© The Author(s), under exclusive license to Springer Nature Switzerland AG 2021   59
B. Gustafsson, Y.-L. Lin, *Laplacian Growth on Branched Riemann Surfaces*,
Lecture Notes in Mathematics 2287, https://doi.org/10.1007/978-3-030-69863-8_5

time a zero of g for the corresponding Löwner-Kufarev equation reaches the unit circle. Unfortunately, as we have not been able to settle Lemma 5.3 (Conjecture), we have to include the validity of this among the assumptions in the theorem below. The precise formulation is as follows.

Theorem 5.1 *Let $f(\cdot, 0) \in \mathcal{O}_{norm}(\overline{\mathbb{D}})$ be given, together with $q(t) \geq 0$ $(0 \leq t < \infty)$ such that $Q(t) \to \infty$ as $t \to \infty$. Then, under the assumption that Lemma 5.3 (Conjecture) is true, there exists a Riemann surface $\mathcal{M}$, a nonconstant holomorphic function $p : \mathcal{M} \to \mathbb{C}$ and a point $\tilde{0} \in \mathcal{M}$ with $p(\tilde{0}) = 0$ such that the following assertions hold.*

(a) *$f(\cdot, 0)$ factorizes over $\mathcal{M}$, i.e. there exists a univalent function $\tilde{f}(\cdot, 0) : \mathbb{D} \to \mathcal{M}$ with $\tilde{f}(0, 0) = \tilde{0}$ such that $f(\cdot, 0) = p(\tilde{f}(\cdot, 0))$.*
(b) *Setting $\tilde{\Omega}(0) = \tilde{f}(\mathbb{D}, 0)$, the weak Hele-Shaw evolution $\{\tilde{\Omega}(t)\}$ on $\mathcal{M}$ defined by*

$$Bal\left(2\pi Q(t)\delta_{\tilde{0}} + \chi_{\tilde{\Omega}(0)}\tilde{m}\right) = \chi_{\tilde{\Omega}(t)}\tilde{m},$$

exists for all $0 \leq t < \infty$, and $\tilde{\Omega}(t)$ is simply connected for each t.
(c) *Let $\nu_{f(\cdot,t)}$ denote the counting function of $f(\cdot, t) = p(\tilde{f}(\cdot, t))$ and let $\Omega(t)$ denote the domain obtained by partial balayage of $\nu_{f(\cdot,t)}m$ onto Lebesgue measure m:*

$$Bal\left(\nu_{f(\cdot,t)}m, m\right) = \chi_{\Omega(t)}m.$$

Then

$$Bal\left(2\pi Q(t)\delta_0 + \chi_{\Omega(0)}m\right) = \chi_{\Omega(t)}m,$$

hence the family $\{\Omega(t)\}$ is a weak solution in the ordinary sense on $\mathbb{C}$, with the domains $\Omega(t)$ possibly multiply connected.

The theorem is pictorially illustrated in Figs. 1.1, 1.2 in Chap. 1, where the vertical time evolution (left track, upwards) represents the main assertion (b) of the theorem, while assertion (c) is represented by the horizontal arrows marked Bal $(\cdot, 1)$.

For the proof of Theorem 5.1 we shall need Lemma 5.1–5.3 below.

Lemma 5.1 *Let $f \in \mathcal{O}_{norm}(\overline{\mathbb{D}})$ and let $0 < r < 1$. Then the following are equivalent.*

(i) *f extends to be meromorphic in $\mathbb{D}(0, \frac{1}{r})$ with poles only at the reflected (in $\partial\mathbb{D}$) zeros of g, more precisely so that $fg^* \in \mathcal{O}(\mathbb{D}(0, \frac{1}{r}) \setminus \overline{\mathbb{D}})$.*
(ii) *For every number ρ with $r < \rho < 1$ there exists a constant C_ρ such that*

$$\left| \int_{\mathbb{D}} h|g|^2 dm \right| \leq C_\rho \sup_{\mathbb{D}(0,\rho)} |h| \quad (h \in \mathcal{O}(\overline{\mathbb{D}})). \tag{5.1}$$

(iii) *For every number ρ with $r < \rho < 1$ there exists a (signed) measure σ with* supp $\sigma \subset \overline{\mathbb{D}}(0, \rho)$ *such that*

$$\int_{\mathbb{D}} h|g|^2 dm = \int_{\mathbb{D}} h d\sigma \quad (h \in \mathcal{O}(\overline{\mathbb{D}})). \tag{5.2}$$

Proof Assume (i). Then for every $r < \rho < 1$ we have

$$\int_{\mathbb{D}} h|g|^2 dm = \frac{1}{2i} \int_{\partial \mathbb{D}} h \bar{f} df = \frac{1}{2i} \int_{\partial \mathbb{D}} h f^* g d\zeta = \frac{1}{2i} \int_{\partial \mathbb{D}(0,\rho)} h f^* g d\zeta,$$

where we used that $f^* g = (fg^*)^* \in \mathcal{O}(\mathbb{D} \setminus \overline{\mathbb{D}}(0, r))$, by assumption. Now (ii) follows with $C_\rho = \frac{1}{2} \int_{\partial \mathbb{D}(0,\rho)} |f^* g||d\zeta|$.

That (ii) implies (iii) follows from general functional analysis, namely the Hahn-Banach theorem and the Riesz representation theorem for functionals on $C(\overline{\mathbb{D}}(0, \rho))$, see [91].

Assume now (iii) and we shall prove (i). Consider the **Cauchy transform** of σ, and that of $|g|^2 \chi_{\mathbb{D}}$, defined by

$$\hat{\sigma}(z) = \frac{1}{\pi} \int_{\mathbb{D}} \frac{d\sigma(\zeta)}{z - \zeta},$$

$$G(z) = \frac{1}{\pi} \int_{\mathbb{D}} \frac{|g(\zeta)|^2 dm(\zeta)}{z - \zeta}, \tag{5.3}$$

respectively. Here G is defined and continuous in all $\mathbb{C}$ and satisfies, in the sense of distributions,

$$\frac{\partial G}{\partial \bar{z}} = \bar{g} g \chi_{\mathbb{D}}.$$

Thus, in $\mathbb{D}$,

$$G = \bar{f} g + H$$

for some $H \in \mathcal{O}(\mathbb{D})$. This equality also defines H on $\partial \mathbb{D}$, by which it becomes continuous on $\overline{\mathbb{D}}$.

On the other hand, (5.2) shows that $\hat{\sigma} = G$ outside $\overline{\mathbb{D}}$, and by continuity this also holds on $\partial \mathbb{D}$. Hence

$$f^* g = \bar{f} g = G - H = \hat{\sigma} - H$$

on $\partial \mathbb{D}$, and since the right member is holomorphic in $\mathbb{D} \setminus \overline{\mathbb{D}}(0, \rho)$ the desired meromorphic extension of f follows. □

If $f(\cdot, t) \in \mathscr{O}_{\mathrm{univ}}(\overline{\mathbb{D}})$ is a univalent weak solution then it is known [36, 47] that the radius of analyticity of f is an increasing function of time. In the non-univalent case this is no longer true, but there is a related radius (essentially $1/r$ in the previous lemma) which is stable in time (actually increases), and this will be a good enough statement for our needs.

Lemma 5.2 *Let* $\tilde{\Omega}(\cdot, t) = \tilde{f}(\mathbb{D}, t)$ *be a simply connected weak solution on a Riemann surface* $\mathscr{M}$ *with projection* $p : \mathscr{M} \to \mathbb{C}$ *and let* $f(\zeta, t) = p(\tilde{f}(\zeta, t))$. *Assume* $q(t) \geq 0$ *and that for some* $0 < r < 1$ *the equivalent conditions in Lemma 5.1 hold for* $f = f(\cdot, 0)$. *Then they hold with the same* r *for all* $f(\cdot, t)$, $t > 0$.

Proof If $f(\cdot, t) \in \mathscr{O}_{\mathrm{norm}}(\overline{\mathbb{D}})$ is a weak solution on $\mathscr{M}$ starting at $t = 0$ then, by (4.28),

$$\int_{\mathbb{D}} h(z)|g(z, t)|^2 \, dm(z) = \int_{\mathbb{D}} h(\varphi(\zeta, 0, t))|g(\zeta, 0)|^2 \, dm(\zeta) + 2\pi \, Q(t) h(0)$$

for all $h \in \mathscr{O}(\overline{\mathbb{D}})$. Recall that $\varphi(\zeta, 0, t) = \tilde{f}^{-1}(\tilde{f}(\zeta, 0), t))$. Assume now that condition (ii) of Lemma 5.1 holds at $t = 0$ for some $0 < r < 1$. Since $|\varphi(\zeta, 0, t)| \leq |\zeta|$ by Schwarz' lemma we then get, for an arbitrary ρ with $r < \rho < 1$,

$$\left| \int_{\mathbb{D}} h(\varphi(\zeta, 0, t))|g(\zeta, 0)|^2 dm(\zeta) \right| \leq C_\rho \sup_{\zeta \in \mathbb{D}(0,\rho)} |h(\varphi(\zeta, 0, t))| \leq C_\rho \sup_{z \in \mathbb{D}(0,\rho)} |h(z)|,$$

hence

$$\left| \int_{\mathbb{D}} h(z)|g(z, t)|^2 \, dm(z) \right| \leq (C_\rho + 2\pi \, Q(t)) \sup_{z \in \mathbb{D}(0,\rho)} |h(z)|.$$

This shows that (ii) of Lemma 5.1 holds also at any time $t > 0$, with the same r as for $t = 0$, which is what we needed to prove.

$\square$

5.2   Statement of Conjecture, and Partial Proofs

The final auxiliary result is a conjecture, which is very likely to be true but for which we have no complete proof at present. It was therefore was listed among the assumptions in Theorem 5.1. It concerns the issue of keeping $\tilde{\Omega}(t)$ simply connected all the time. With the weak solution pulled back to $\mathbb{D}$ as in Example 4.1 (specifically Eq. (4.16) there) the crucial statement becomes the following.

Lemma 5.3 (Conjecture) *Let* $g \in \mathcal{O}(\overline{\mathbb{D}})$ *be fixed (independent of t) and denote by* $\{D(t) : 0 \leq t < \varepsilon\}$ *the weak solution for the weight* $|g|^2$ *and initial domain* $D(0) = \mathbb{D}$, *with* $\varepsilon > 0$ *chosen so small that all domains* $D(t)$ *are compactly contained in the region of analyticity of g. Thus,*

$$\int_{D(t)} h|g|^2 dm \geq \int_{\mathbb{D}} h|g|^2 dm + 2\pi Q(t)h(0)$$

for every $h \in SL^1(D(t), |g|^2 m)$, *and* $0 \leq t \leq \varepsilon$. *Then, if* $\varepsilon > 0$ *is sufficiently small, the domains* $D(t)$ *are all simply connected.*

By assumption, g is analytic in some disk $\mathbb{D}_R$ with $R > 1$. In terms of partial balayage, the domain $D(t)$ (with $\mathbb{D} \subset D(t) \subset \mathbb{D}_R$) is then given by

$$\mathrm{Bal}\,(2\pi Q(t)\delta_0, |g|^2 \chi_{\mathbb{D}_R \setminus \mathbb{D}}) = |g|^2 \chi_{D(t) \setminus \mathbb{D}}.$$

If g has no zeros on $\partial \mathbb{D}$, then stability results for free boundaries [12, 23], alternatively [107], imply that the domains $D(t)$ are indeed simply connected. This argument was used also in Proposition 5.1 of [100] for the same conclusion. One can also use existence of classical solutions [19]. However, we need Lemma 5.3 exactly in the case when g has zeros on $\partial \mathbb{D}$.

It is natural to conjecture that the $D(t)$ are not only simply connected, but also starshaped with respect to the origin, for $t > 0$ small enough. Such a statement has the advantage of being quantitative rather than just topological.

Below we outline some steps towards a proof (so far incomplete) of starshapedness of the $D(t)$. In terms of the function $u = u(z, t)$ appearing in Definition 3.2 for the choice $\mu = 2\pi Q(t)\delta_0$, $\lambda = |g|^2 \chi_{\mathbb{D}_R \setminus \mathbb{D}}$, the weak solution $\{D(t) : 0 \leq t < \varepsilon\}$ is given by

$$D(t) = \{z \in \mathbb{C} : u(z, t) > 0\}$$

with u satisfying (and determined by)

$$\begin{cases} u \geq 0 \quad \text{in} \quad \mathbb{C}, \\ \Delta u = |g|^2 \chi_{D(t) \setminus \mathbb{D}} - 2\pi Q(t)\delta_0 \quad \text{in} \quad \mathbb{C}, \\ u = |\nabla u| = 0 \quad \text{outside} \quad D(t). \end{cases} \tag{5.4}$$

These properties follow in a standard manner from the complementarity system (of type (3.8), (3.9)) satisfied by u.

Now write, in terms of polar coordinates $z = re^{i\theta}$,

$$v = r\frac{\partial u}{\partial r}.$$

This function is continuous in $\mathbb{C} \setminus \{0\}$ because the elliptic partial differential equation which u satisfies (in the sense of distributions) shows that u is continuously differentiable, even across $\partial D(t)$. In order to show that $D(t)$ is star-shaped it is enough to show that u decreases in each radial direction, that is, that $v \leq 0$. This is what one hopes to show, for $t > 0$ small enough.

The equation satisfied by u takes in polar coordinates the form

$$\frac{1}{r}\frac{\partial}{\partial r}(r\frac{\partial u}{\partial r}) + \frac{1}{r^2}\frac{\partial^2 u}{\partial \theta^2} = |g|^2 \chi_{D(t)\setminus\mathbb{D}} - 2\pi Q(t)\delta_0 \quad \text{in} \quad \mathbb{C}.$$

Multiplying both members with r^2 and then acting by $\frac{1}{r}\frac{\partial}{\partial r}$ shows that v satisfies

$$\Delta v = \frac{1}{r}\frac{\partial}{\partial r}(r^2|g|^2\chi_{D(t)\setminus\mathbb{D}}) \quad \text{in} \quad \mathbb{C}. \tag{5.5}$$

This is to be interpreted in the sense of distributions. Note that the Dirac distribution in the right member of the previous equation has disappeared. At the origin, $u(z) = -Q(t)\log|z| + \text{harmonic}$, and except for this logarithmic singularity u is continuously differentiable in all $D(t)$. It follows that v is harmonic in $\mathbb{D}$ with $v(0) = -Q(t)$ and that v is continuous in all $D(t)$, in fact Lipschitz continuous by [23] (Theorem 4 in Chapter 1).

There is a distributional contribution in the right member of (5.5) on $\partial\mathbb{D}$, namely equal to $|g|^2$ times arc length measure on $\partial\mathbb{D}$. In particular, this contribution is nonnegative. There is also a distributional contribution on $\partial D(t)$, but we shall rather consider this set as the outer boundary for the region of consideration. Since u vanishes together with its first derivative on $\partial D(t)$, the boundary condition for v there is

$$v = 0 \quad \text{on} \quad \partial D(t).$$

Inside $D(t) \setminus \mathbb{D}$, Eq. (5.5) can be written

$$\Delta v = 2|g|^2\text{Re}(1 + \frac{zg'}{g}), \quad z \in D(t) \setminus \mathbb{D}. \tag{5.6}$$

Unfortunately we cannot be sure of the sign of the right member of (5.6). If we knew that it was nonnegative, then the desired conclusion $v \leq 0$ would follow immediately from the maximum principle. To clarify the situation let us make a local analysis around a point on $\partial\mathbb{D}$ at which g vanishes. We may assume that this point is $z = 1$, and then write

$$g(z) = (z - 1)^d h(z),$$

where d is the order of the zero and h is analytic with $h(1) \neq 0$. We compute the right member of (5.6) as

$$2|g(z)|^2 \operatorname{Re}(1 + \frac{zg'(z)}{g(z)}) = 2|z - 1|^{2d}|h(z)|^2 \operatorname{Re}(1 + \frac{dz}{z-1} + \frac{zh'(z)}{h(z)})$$

$$= 2|z - 1|^{2d} \operatorname{Re}(zh'(z)\overline{h(z)}) + 2|z - 1|^{2(d-1)}|h(z)|^2 \operatorname{Re}((z-1)(\bar{z} - 1) + d \cdot z(\bar{z} - 1))$$

$$= 2|z - 1|^{2(d-1)}|h(z)|^2 \left(|z - 1|^2 \operatorname{Re}\frac{zh'(z)}{h(z)} + (d+1)\left(|z - \frac{d+2}{2d+2}|^2 - (\frac{d}{2d+2})^2\right)\right).$$

Here the second term inside the bracket is positive outside the circle with center $\frac{d+2}{2d+2}$ and radius $\frac{d}{2d+2}$, in particular outside $\mathbb{D}$, while the first term may have any sign. For $d = 1$ the above expression takes the simpler form

$$2|g(z)|^2 \operatorname{Re}(1 + \frac{zg'(z)}{g(z)}) = 2|h(z)|^2 \left(|z - 1|^2 \operatorname{Re}\frac{zh'(z)}{h(z)} + 2\left(|z - \frac{3}{4}|^2 - (\frac{1}{4})^2\right)\right).$$

In any case, the right member in (5.6) is positive in major parts of neighborhoods (outside $\mathbb{D}$) of points on $\partial\mathbb{D}$ where g vanishes. In some examples, like if h is constant, which will be the case in the example in Sect. 7.1.3 below, it follows that the right member in (5.6) is positive in all $D(t) \setminus \mathbb{D}$, and the star-shapedness can be inferred. Close to other points on $\partial\mathbb{D}$ one can perform an analysis based on known stability behavior of free boundaries [12, 23]. This gives at least that, outside any fixed neighborhood of $z = 1$, $D(t)$ does not have any holes if $t > 0$ is small enough.

This finishes the discussion of steps towards showing that the $D(t)$ are star-shaped.

Proof of Theorem 5.1 To get started, observe that we can find $\mathcal{M}$ so that (a) holds. It is just to take $\mathcal{M} = \mathbb{D}$, $\tilde{f}(\zeta, 0) = \zeta$ and $p = f(\cdot, 0)$. Compare the proof of Lemma 2.1. The remaining part of the proof consists of extending the Riemann surface $\mathcal{M}$ so that the weak solution in (b) exists and is simply connected for gradually increasing values of t.

So assume that we have constructed $\mathcal{M}$ so that (b) holds on some time interval $[0, T]$, where $T \geq 0$ ($T = 0$ not excluded). Since the solution is kept simply connected also the conformal map $\tilde{f}(\cdot, t)$ is automatically extended, to times $t \in [0, T]$. The part $\mathcal{M}_T = \tilde{\Omega}(T) = \tilde{f}(\mathbb{D}, T) \subset \mathcal{M}$ of the extended surface will then be permanent and will no more be changed. However, it may happen that we will have to modify $\mathcal{M} \setminus \mathcal{M}_T$ in some future time $t > T$ by introducing more branch points. Therefore, to make the procedure consistent, we throw away $\mathcal{M} \setminus \mathcal{M}_T$ and keep only $\mathcal{M}_T = \tilde{\Omega}(T)$ (for the time being).

We have that $\tilde{\Omega}(t) = \tilde{f}(\mathbb{D}, t)$ for all $0 \leq t \leq T$, $f(\cdot, t) = p \circ \tilde{f}(\cdot, t) \in \mathcal{O}_{\mathrm{norm}}(\overline{\mathbb{D}})$, $f(\cdot, t)$ is a subordination chain and $f(\cdot, t)$ is meromorphic in a disk $\mathbb{D}(0, \frac{1}{r})$, with $0 < r < 1$ independent of t by Lemma 5.2. Now choose $1 < \rho < \frac{1}{r}$ with r as in Lemma 5.2, so that $f'(\zeta, T)$ has no zeros for $1 < |\zeta| < \rho$ (but may have it for $|\zeta| = 1$). Viewing $\mathbb{D}(0, \rho) \supset \mathbb{D}$ as a Riemann surface over $\mathbb{C}$ with covering map $f(\cdot, T)$ we get, on the level of abstract Riemann surfaces, a (new) preliminary extension $\mathcal{M} \supset \mathcal{M}_T$ of $\mathcal{M}_T$. On $\mathcal{M}$ we can continue the weak solution to some time interval $[0, T + \varepsilon]$, $\varepsilon > 0$. Compare the discussion in Example 4.1. For $\varepsilon > 0$ small enough this solution $\tilde{\Omega}(t)$ remains simply connected, assuming Lemma 5.3. Then set $\mathcal{M}_{T+\varepsilon} = \tilde{\Omega}(T + \varepsilon)$.

Thus we can always extend a weak solution defined on a closed time interval $[0, T]$ to some larger interval $[0, T + \varepsilon]$. We also have to show that whenever we have a solution on a half-open interval $[0, T)$ (with $T > 0$) it can be extended to the closure $[0, T]$. However, this is fairly immediate because we can simply define $\tilde{\Omega}(T) = \mathcal{M}_T = \cup_{0 \leq t < T} \mathcal{M}_t$ (compare proof of Lemma 2.1). This surface is easily seen to be simply connected (because any closed curve in $\mathcal{M}_T$ will lie entirely in $\mathcal{M}_t$ for some $t < T$). Moreover, the defining property (4.11) of a weak solution will hold on all $[0, T]$, and since the radius of analyticity of $f(\cdot, T) : \mathbb{D} \to \tilde{\Omega}(T)$ is larger than one (Lemma 5.2), $\tilde{\Omega}(T) \cong \mathbb{D}$ will have compact closure in a larger Riemann surface $\mathcal{M} \supset \mathcal{M}_T$, on which the evolution may continue.

The above arguments show that there is no finite stopping time for the construction of $\mathcal{M}$ and a simply connected weak solution in $\mathcal{M}$. Therefore part (b) of the theorem follows.

Assertion (c) of the theorem is an easy consequence of Proposition 4.3.   □

5.3 Discussion

Much related to our results is the work [53] by H. Hedenmalm and S. Shimorin. One of their main results says that if a planar domain $\mathcal{M} \subset \mathbb{C}$ is provided with a hyperbolic metric

$$ds^2 = \lambda^2(dx^2 + dy^2),$$

then the weak Hele-Shaw evolution $t \mapsto \Omega(t)$ starting from the empty set ($\Omega(0) = \emptyset$) and with injection at one point $a \in \mathcal{M}$ consists of simply connected domains as long as the solution exists. In our balayage language the domains $\Omega(t)$ are given by

$$\mathrm{Bal}\,(t\delta_a, \lambda^2) = \lambda^2 \chi_{\Omega(t)},$$

and the hyperbolicity of the metric means that the Gaussian curvature is non-positive,

$$\kappa = -\frac{\Delta \log \lambda}{\lambda^2} \leq 0,$$

hence that $\log \lambda$ is subharmonic. The condition of hyperbolicity can actually be somewhat relaxed, see [51, 52].

It is assumed in [53] that λ is strictly positive and real analytic. If the manifold $\mathscr{M}$ is complete with respect to the metric then the solution exists for all time $0 < t < \infty$ and the domains $\Omega(t)$ exhaust $\mathscr{M}$. In that case the solution can be used to define a kind of global polar coordinates on $\mathscr{M}$, equivalently to define what the authors call a **Hele-Shaw exponential mapping**. This is an orthogonal system of coordinates, one coordinate being the time t at which the Hele-Shaw front $\partial\Omega(t)$ reaches the point in question and the other coordinate parametrizing the individual fluid particles on $\partial\Omega(t)$.

It is shown also in [53] that the solution is real analytic with respect to time. Even more, at time $t = 0$ the solution can be extended real-analytically in the negative direction, to some interval $-\varepsilon < t < 0$ for $\varepsilon > 0$ sufficiently small. This reminds a little of what we shall encounter in Chap. 7, where the cardioid solution in Sect. 7.1.3 is extended backwards in Sect. 7.1.4.

In our present work we have Hele-Shaw evolution on a degenerate hyperbolic manifold: the metric is given by $\lambda = |g| = |f'|$, see (4.1), for which

$$\Delta \log \lambda = 2\pi \sum_j \delta_{\omega_j},$$

where ω_j are the zeros of g. Hence the curvature is concentrated to the branch points, where it has singular negative contributions. The Gaussian curvature κ does not really make sense at the branch points, but one may consider directly the curvature matrix $\Theta = -\frac{1}{2}(\Delta \log \lambda)dz \wedge d\bar{z}$ (see [27]), which has (negative) point charges at the branch points.

The proof in [53] that the fluid domains are simply connected depends crucially on the solution starting from the empty set. In our case, when starting with an arbitrary simply connected domain $\Omega(0)$, the solution will in general not stay simply connected all the time. This is exactly the reason why we have to let it go up on covering surfaces, when simple connectivity is threatened. The arguments in [53] involve applying the Laplacian operator a second time on functions u which appear in systems like (5.4). This gives, after some suitable arrangements made in [53], that u can be identified with the fundamental solution of the biharmonic equation. After pull-back to the unit disk and comparing with the explicit fundamental solution there, the authors obtain estimates which guarantee simple connectedness. Results of M. Sakai also play a role in the mentioned arguments. As a whole, the proof is highly nontrivial, and it does not directly carry over to our case.

Chapter 6
General Structure of Rational Solutions

Abstract The behaviour of such Riemann surface weak solutions for which the derivative of the conformal map is a rational function is investigated. The solution stays on the same form locally, as long as the zeros of the mentioned derivative stay away from the unit circle. However, when such zeros do reach the unit circle and new branch points are created, the rational function changes its structure, acquiring new zeros and poles. Two approaches for handling these questions are used: one direct approach and one via quadrature Riemann surfaces.

6.1 Introduction

In this section we shall prove that the property of $g = f'$ being a rational function is preserved in time for weak solutions as long as they remain simply connected. In other words, it is preserved by the Löwner-Kufarev equation, even under transition of zeros of g through $\partial \mathbb{D}$. However, g acquires additional zeros and poles under such an event, and the transition will not be smooth. We shall also show that for certain other solutions of the Polubarinova-Galin equation rationality is also preserved.

We shall give two avenues to the question of rationality: first a direct approach, just making an 'Ansatz' of a rational g in a suitable version of the Polubarinova-Galin equation, and secondly via quadrature identities, which are related to the concept of a weak solution and which can incorporate transitions of zeros through $\partial \mathbb{D}$.

6.2 Direct Approach

We assume that g rational of the form (2.15), and we address the question to which extent this form is preserved in time for solutions of the Polubarinova-Galin or Löwner-Kufarev equations when g is allowed to have zeros in $\mathbb{D}$.

© The Author(s), under exclusive license to Springer Nature Switzerland AG 2021
B. Gustafsson, Y.-L. Lin, *Laplacian Growth on Branched Riemann Surfaces*,
Lecture Notes in Mathematics 2287, https://doi.org/10.1007/978-3-030-69863-8_6

Recall (Theorem 2.2) that the Polubarinova-Galin equation is equivalent to a relaxed version of the Löwner-Kufarev equation,

$$\dot{f}(\zeta, t) = \zeta f'(\zeta, t) (P(\zeta, t) + R(\zeta, t)), \tag{6.1}$$

where $P(\zeta, t)$ is the Poisson integral (2.11) and where $R(\zeta, t)$ is any function of the form (2.27). We shall here assume, for simplicity, that the zeros $\omega_j \in \mathbb{D}$ of g are simple, and then (2.27) becomes

$$R(\zeta, t) = \mathrm{i}\,\mathrm{Im}\sum_{\omega_j \in \mathbb{D}} \frac{2B_j(t)}{\omega_j(t)} + \sum_{\omega_j \in \mathbb{D}} \left(\frac{2B_j(t)}{\zeta - \omega_j(t)} - \frac{2\overline{B_j(t)}\zeta}{1 - \overline{\omega_j(t)}\zeta} \right). \tag{6.2}$$

The interpretation of the free constants $B_j(t)$ is that they determine the speed of the branch points $f(\omega_j(t), t)$. Indeed, using (6.2) we have

$$\frac{d}{dt} f(\omega_j(t), t) = 2\omega_j(t) f''(\omega_j(t), t) B_j(t).$$

Equation (6.1) for f is equivalent to the equation, generalizing (2.16),

$$\frac{\partial}{\partial t}(\log g) = \zeta(P(\zeta, t) + R(\zeta, t))\frac{\partial}{\partial \zeta}(\log g) + \frac{\partial}{\partial \zeta}(\zeta(P(\zeta, t) + R(\zeta, t))) \tag{6.3}$$

for $g = g(\zeta, t)$. Here the derivatives of $\log g$ are obtained from (2.18), (2.19), and it only remains to evaluate the Poisson integral $P(\zeta, t)$. This can be done by a simple residue calculus in (2.11) using that $|g(\zeta, t)|^2 = g(\zeta, t)g^*(\zeta, t)$ when $\zeta \in \partial\mathbb{D}$, where the right member is a rational function in ζ and where we recall from (2.8) that g^* denotes the holomorphic reflection of a function g in $\partial\mathbb{D}$. However, the calculation becomes more transparent if everything is done at an algebraic level, by which it essentially reduces to an expansion in partial fractions.

Recall that, by definitions of P and R,

$$P(\zeta, t) + P^*(\zeta, t) = \frac{2q(t)}{g(\zeta, t)g^*(\zeta, t)},$$

$$R(\zeta, t) + R^*(\zeta, t) = 0. \tag{6.4}$$

The rational function $q(t)/g(\zeta, t)g^*(\zeta, t)$ has poles at the zeros of g and g^*, i.e. at $\omega_1, \ldots, \omega_m, \omega_1^*, \ldots, \omega_m^*$ (here $\omega_j^* = 1/\bar{\omega}_j$). At infinity it has the behavior (by (2.15))

$$\lim_{\zeta \to \infty} \frac{q(t)}{g(\zeta, t)g^*(\zeta, t)} = A_\infty = \begin{cases} \dfrac{q \prod_{j=1}^{n} \bar{\zeta}_j}{|b|^2 \prod_{j=1}^{m} \bar{\omega}_j} & \text{if } m = n, \\[2ex] 0 & \text{if } m > n. \end{cases} \tag{6.5}$$

We shall assume, in addition to the zeros ω_k being simple, that no two zeros are reflections of each other with respect to the unit circle, i.e. we assume that $\omega_k \neq \omega_j^*$ for all k, j and, in particular $(k = j)$, that there are no zeros on the unit circle. These assumptions are necessary in order to expect existence of a smooth solution of the Polubarinova-Galin equation (2.1), and even more so for the Löwner-Kufarev equation. Indeed, spelling out (2.1) as

$$\dot{f}(\zeta, t) \cdot \zeta^{-1} g^*(\zeta, t) + \dot{f}^*(\zeta, t) \cdot \zeta g(\zeta, t) = 2q(t) \qquad (6.6)$$

we see that if, for some particular value of t, g and g^* have a common zero, with $\dot{f}$ and $\dot{f}^*$ finite, then $q(t)$ must be zero.

It will be seen shortly (Eq. 6.9) that P is a rational function whenever g is rational, hence, by (6.1), also $\dot{f}$ is rational. So (6.6) is an identity between rational functions and so is valid throughout the Riemann sphere.

With the mentioned assumptions in force we can write

$$\frac{q(t)}{g(\zeta, t)g^*(\zeta, t)} = A_\infty + \sum_{k=1}^{m} \frac{\overline{A_k}}{\overline{\omega_k}} + \sum_{k=1}^{m} \left[\frac{A_k}{\zeta - \omega_k} + \frac{\overline{A_k}\zeta}{1 - \overline{\omega_k}\zeta} \right], \qquad (6.7)$$

where the coefficients $A_k = A_k(t, b, \omega_1, \ldots, \omega_m, \zeta_1, \ldots, \zeta_n)$ are given by

$$A_k = \frac{q(t)}{g'(\omega_k, t)g^*(\omega_k, t)} = \frac{q}{|b|^2} \cdot \frac{\prod_j(\omega_k - \zeta_j) \prod_j \overline{(\omega_k^* - \zeta_j)}}{\prod_{j \neq k}(\omega_k - \omega_j) \prod_j \overline{(\omega_k^* - \omega_j)}} \qquad (6.8)$$

for $1 \leq k \leq m$. Notice that some of the A_k may vanish: if $\omega_k \in \mathbb{D}$ and ω_k^* coincides with one of the poles ζ_j, then $A_k = 0$.

Now, $P(\zeta, t)$ is to be that holomorphic function in $\mathbb{D}$ whose real part has boundary values $q(t)/g(\zeta, t)g^*(\zeta, t)$ and whose imaginary part vanishes at the origin. The function (6.7) itself certainly has the right boundary behaviour on $\partial\mathbb{D}$, but it is not holomorphic in $\mathbb{D}$. On the other hand, the two types of polar parts occurring in (6.7) have the same real parts on the boundary:

$$\text{Re} \, \frac{A_k}{\zeta - \omega_k} = \text{Re} \, \frac{\overline{A_k}\zeta}{1 - \overline{\omega_k}\zeta} \quad \text{on } \partial\mathbb{D}.$$

Therefore, without changing the real part on the boundary we can make the function (6.7) holomorphic in $\mathbb{D}$ by a simple exchange of polar parts. In addition, one can add a purely imaginary constant to account for the normalization of P at the origin. This gives

$$P(\zeta, t) = A_0 + \sum_{\omega_j \in \mathbb{C} \setminus \overline{\mathbb{D}}} \frac{2A_j}{\zeta - \omega_j} + \sum_{\omega_j \in \mathbb{D}} \frac{2\overline{A_j}\zeta}{1 - \overline{\omega_j}\zeta}, \qquad (6.9)$$

with the $A_j = A_j(t)$ given by (6.8) for $1 \le j \le m$. For A_0 we have

$$\operatorname{Re} A_0 = \operatorname{Re} A_\infty + \operatorname{Re} \sum_{k=1}^{m} \frac{A_k}{\omega_k},$$

$$\operatorname{Im} A_0 = \operatorname{Im} \sum_{\omega_j \in \mathbb{C} \backslash \overline{\mathbb{D}}} \frac{2A_j}{\omega_j},$$

so that the real part of (6.7) remains unaffected in the passage to (6.9), and so that the normalization $\operatorname{Im} P(0,t) = 0$ is achieved. Note that $\operatorname{Re}(P(\zeta,t) + R(\zeta,t)) \ge 0$ in $\mathbb{D}$ if and only if $R = 0$ (because R has poles in $\mathbb{D}$ if $R \ne 0$).

Thus

$$P(\zeta,t) + R(\zeta,t)$$

$$= A_0 + i \operatorname{Im} \sum_{\omega_j \in \mathbb{D}} \frac{2B_j}{\omega_j} + \sum_{\omega_j \in \mathbb{C} \backslash \overline{\mathbb{D}}} \frac{2A_j}{\zeta - \omega_j} + \sum_{\omega_j \in \mathbb{D}} \frac{2B_j}{\zeta - \omega_j} + \sum_{\omega_j \in \mathbb{D}} \frac{2(\overline{A}_j - \overline{B}_j)\zeta}{1 - \overline{\omega}_j \zeta}$$

$$= C + \sum_{\omega_j \in \mathbb{C} \backslash \overline{\mathbb{D}}} \frac{2A_j}{\zeta - \omega_j} + \sum_{\omega_j \in \mathbb{D}} \frac{2B_j}{\zeta - \omega_j} - \sum_{\omega_j \in \mathbb{D}} \frac{2(\overline{A}_j - \overline{B}_j)(\omega_j^*)^2}{\zeta - \omega_j^*},$$

where

$$C = A_0 + i \operatorname{Im} \sum_{\omega_j \in \mathbb{D}} \frac{2B_j}{\omega_j} - \sum_{\omega_j \in \mathbb{D}} 2(\overline{A}_j - \overline{B}_j)\omega_j^*.$$

Also,

$$\zeta(P(\zeta,t) + R(\zeta,t))$$

$$= C\zeta + D + \sum_{\omega_j \in \mathbb{C} \backslash \overline{\mathbb{D}}} \frac{2A_j \omega_j}{\zeta - \omega_j} + \sum_{\omega_j \in \mathbb{D}} \frac{2B_j \omega_j}{\zeta - \omega_j} - \sum_{\omega_j \in \mathbb{D}} \frac{2(\overline{A}_j - \overline{B}_j)(\omega_j^*)^3}{\zeta - \omega_j^*},$$

with

$$D = \sum_{\omega_j \in \mathbb{C} \backslash \overline{\mathbb{D}}} 2A_j + \sum_{\omega_j \in \mathbb{D}} 2B_j - \sum_{\omega_j \in \mathbb{D}} 2(\overline{A}_j - \overline{B}_j)(\omega_j^*)^2.$$

In view of (2.18), (2.19) the dynamical law (6.3) becomes

$$
\frac{\dot{b}}{b} - \sum_{k=1}^{m} \frac{\dot{\omega}_k}{\zeta - \omega_k} + \sum_{j=1}^{n} \frac{\dot{\zeta}_j}{\zeta - \zeta_j}
\tag{6.10}
$$

$$
= \left(C\zeta + D + \sum_{\omega_j \in \mathbb{C} \backslash \overline{\mathbb{D}}} \frac{2A_j \omega_j}{\zeta - \omega_j} + \sum_{\omega_j \in \mathbb{D}} \frac{2B_j \omega_j}{\zeta - \omega_j} - \sum_{\omega_j \in \mathbb{D}} \frac{2(\overline{A}_j - \overline{B}_j)(\omega_j^*)^3}{\zeta - \omega_j^*} \right)
$$

$$
\cdot \left(\sum_{k=1}^{m} \frac{1}{\zeta - \omega_k} - \sum_{j=1}^{n} \frac{1}{\zeta - \zeta_j} \right)
$$

$$
+ C - \sum_{\omega_j \in \mathbb{C} \backslash \overline{\mathbb{D}}} \frac{2A_j \omega_j}{(\zeta - \omega_j)^2} - \sum_{\omega_j \in \mathbb{D}} \frac{2B_j \omega_j}{(\zeta - \omega_j)^2} + \sum_{\omega_j \in \mathbb{D}} \frac{2(\overline{A}_j - \overline{B}_j)(\omega_j^*)^3}{(\zeta - \omega_j^*)^2}. \tag{6.11}
$$

The derivatives $\dot{b}$, $\dot{\omega}_k$, $\dot{\zeta}_j$ to be determined appear as coefficients in the constant term and poles of order one. Therefore (6.10) can be satisfied only if all terms with poles of higher order cancel. This automatically occurs for the terms of the form $\frac{2A_j \omega_j}{(\zeta - \omega_j)^2}$ ($\omega_j \in \mathbb{C} \backslash \overline{\mathbb{D}}$) and $\frac{2B_j \omega_j}{(\zeta - \omega_j)^2}$ ($\omega_j \in \mathbb{D}$).

In order that the remaining terms $\sum_{\omega_j \in \mathbb{D}} \frac{2(\overline{A}_j - \overline{B}_j)(\omega_j^*)^3}{(\zeta - \omega_j^*)^2}$ with poles of the second order shall disappear we must have, for each j with $\omega_j \in \mathbb{D}$, that

$$
A_j = B_j. \tag{6.12}
$$

The second order poles cannot cancel in any other way. In order to allow the rational form (2.15) to be stable in time we therefore assume from now on that (6.12) holds.

Note that R under this assumption becomes uniquely determined. When (6.12) holds,

$$
C = A_0 + i \operatorname{Im} \sum_{\omega_j \in \mathbb{D}} \frac{2A_j}{\omega_j} = \operatorname{Re} A_\infty + \operatorname{Re} \sum_{k=1}^{m} \frac{A_k}{\omega_k} + i \operatorname{Im} \sum_{j=1}^{m} \frac{2A_j}{\omega_j}, \tag{6.13}
$$

$$
D = \sum_{j=1}^{m} 2A_j, \tag{6.14}
$$

and $P + R$ takes the simpler form

$$
P(\zeta, t) + R(\zeta, t) = C + \sum_{j=1}^{m} \frac{2A_j}{\zeta - \omega_j}.
$$

The dynamical law (6.10) now becomes

$$
\frac{\dot{b}}{b} - \sum_{k=1}^{m} \frac{\dot{\omega}_k}{\zeta - \omega_k} + \sum_{j=1}^{n} \frac{\dot{\zeta}_j}{\zeta - \zeta_j}
$$

$$
= \left(C\zeta + D + \sum_{j=1}^{m} \frac{2A_j\omega_j}{\zeta - \omega_j} \right) \cdot \left(\sum_{k=1}^{m} \frac{1}{\zeta - \omega_k} - \sum_{j=1}^{n} \frac{1}{\zeta - \zeta_j} \right) + C - \sum_{j=1}^{m} \frac{2A_j\omega_j}{(\zeta - \omega_j)^2}.
$$

$$(6.15)$$

Therefore (6.10) results in the following system of ordinary differential equations for ω_k, ζ_j, b.

Theorem 6.1 *Under the assumption that g has only simple zeros, that g and g^* have no common zeros (in particular g has no zero on $\partial\mathbb{D}$), and that in addition (6.12) holds, the Polubarinova-Galin equation (6.1), or (6.3), gives the following rational dynamics for g:*

$$
\frac{d}{dt} \log \omega_k = -C - \frac{2A_k}{\omega_k} - \sum_{j=1,\, j\neq k}^{m} \frac{2(A_k + A_j)}{\omega_k - \omega_j} + \sum_{j=1}^{n} \frac{2A_k}{\omega_k - \zeta_j}
$$

$$
= P^*(\omega_k) + R^*(\omega_k) - \frac{2A_k}{\omega_k}\left(1 + \sum_{j=1}^{m} \frac{1}{1 - \overline{\omega}_j\omega_k} - \sum_{j=1}^{n} \frac{1}{1 - \overline{\zeta}_j\omega_k}\right), \quad (6.16)
$$

$$
\frac{d}{dt} \log \zeta_j = -C - \sum_{k=1}^{m} \frac{2A_k}{\zeta_j - \omega_k} = P^*(\zeta_j) + R^*(\zeta_j), \tag{6.17}
$$

$$
\frac{d}{dt} \log b = (m - n + 1)C. \tag{6.18}
$$

Here the coefficients A_j, C are given by (6.5), (6.8), (6.13).

Since the $B_j(t)$ are completely free functions one can simply define them by (6.12). Then (6.16)–(6.18) is a regular system of ordinary differential equations, having a unique solution as long as the stated conditions on the zeros of g and g^* hold.

The above unique rational solution of the Polubariova-Galin equation solves the Löwner-Kufarev equation if and only if $R = 0$. By (6.12) this requires that $A_k = 0$ for each k with $\omega_k \in \mathbb{D}$. Looking at (6.8) we see that, if $q \neq 0$, the only way that A_k can vanish at a given instant is that $\omega_k^* = \zeta_j$ for some j. However, it will be seen in Example 6.2 below that A_k may vanish for some particular value of t without vanishing for all t. Therefore we also need that

$$
\dot{\omega}_k^* = \dot{\zeta}_j,
$$

so that the condition $A_k = 0$ persists in time. We shall show that this is the case if and only if ω_k^* is a pole of g of order at least two (more generally, of strictly higher order than that of the zero).

Assume that at one particular moment, say $t = 0$, $\omega_k^* = \zeta_j$ for some pair k, j. Then $A_k = 0$ at that moment so that (6.16), (6.17) (with $R = 0$) become

$$\frac{d}{dt} \log \omega_k = P^*(\omega_k), \tag{6.19}$$

$$\frac{d}{dt} \log \zeta_j = P^*(\zeta_j). \tag{6.20}$$

This can also be written

$$\frac{d}{dt} \log \omega_k^* = -P(\omega_k^*),$$

$$\frac{d}{dt} \log \zeta_j = P^*(\omega_k^*).$$

Thus we see that $\frac{d}{dt} \log \omega_k^* = \frac{d}{dt} \log \zeta_j$ holds if and only if $P(\omega_k^*) + P^*(\omega_k^*) = 0$, which, recalling (6.4), happens if and only if gg^* has a pole at $\omega_k^* = \zeta_j$. Looking at the expression (2.15) for g one sees that this occurs if and only if the pole of g at ζ_j is of higher order than the zero of g at ω_k.

Finally, an informal remark about multiple zeros (a situation which is not covered by the above analysis). If a zero $\omega_k \in \mathbb{C} \setminus \overline{\mathbb{D}}$ is of order ≥ 2 then P will have a pole at ω_k of the same order, and it is easy to see that this pole will never cancel in the dynamical equation (6.10). For this reason multiple zeros outside $\overline{\mathbb{D}}$ can never survive, even though collisions may occur. The solution will remain smooth over a collision because if two roots, ω_1 and ω_2, collide the equations still will be regular when reformulated in terms of the combinations $\omega_1 + \omega_2$ and $\omega_1 \omega_2$.

On the other hand, if g has a multiple zero ω_k in $\mathbb{D}$ this will not cause any higher order pole of P if g has a pole of at least the same order at ω_k^*, and if the order of the pole is of strictly higher order then the same situation will persist in time. Therefore a solution will be obtained as before.

By now we have proved the following theorem on local behavior of solutions of (6.1), or (6.3).

Theorem 6.2 *If $g(\zeta, 0)$ is given on the form (2.15) and such that no two zeros of $g(\zeta, 0)$ are related by $\omega_k = \omega_j^*$, then for exactly one choice of $R(\zeta, t)$, namely that defined by (6.12), there exists a solution $g(\zeta, t)$ of (6.3) which remains on the original rational form (2.15).*

Necessary and sufficient condition for this rational solution to also solve the Löwner-Kufarev equation (2.10) is that $R(\zeta, t) = 0$. This occurs precisely under the condition that whenever $g(\zeta, t)$ has a zero ω_k in $\mathbb{D}$, the reflected point ω_k^* is a pole of $g(\zeta, t)$ of order strictly greater than that of the zero. This property is conserved in time.

Every pole ζ_j of g moves out from the origin, and every zero ω_k inside the unit disk moves towards the origin, as time increases.

The last statement follows from (6.19) together with the fact that P is positive in $\mathbb{D}$, negative outside, and the opposite for P^*.

Remark 6.1 A particular consequence of Theorem 6.2 is that the there are no polynomial solutions of the Löwner-Kufarev equation with zeros of g in the unit disk.

Example 6.1 Consider g being of the form

$$g(\zeta, t) = b(t)(\zeta - \omega_1(t)),$$

with $\omega_1(t) > 0$ and $b(t)$ real. In the notations of (2.15) and Theorem 6.1 we have $m = 1, n = 0, A_\infty = 0, A_1 = \frac{q\omega_1}{b^2(1-|\omega_1|^2)}, C = \frac{A_1}{\omega_1}$. This gives the dynamical system

$$\dot{\omega}_1 = -\frac{3q\omega_1}{b^2(1 - \omega_1^2)}, \tag{6.21}$$

$$\dot{b} = \frac{2q}{b(1 - \omega_1^2)}, \tag{6.22}$$

expressing the necessary and sufficient conditions for (6.1) or (6.3), that is, the Polubarinova-Galin equation (2.1), to hold.

Starting with $b(0) = -1, \omega_1(0) = 1$, and choosing $q(t) = e^{2t} - e^{-4t}$, which is positive for $t > 0$, one verifies that $\omega_1(t) = e^{3t}, b(t) = -e^{-2t}$ is a solution of (6.21), (6.22). The corresponding mapping function $f(\zeta, t)$ is then univalent, and starts out from $f(\zeta, 0) = \zeta - \frac{1}{2}\zeta^2$, which maps $\mathbb{D}$ onto a cardioid with a cusp on the boundary. To be more precise, the system (6.21), (6.22) gives the impression of being singular for the initial location $\omega_1 = 1$ of the zero of g. However, the data above are chosen so that q vanishes at the same time. Indeed, $\frac{q(t)}{1-\omega_1^2(t)} = -e^{-4t}$, which is smooth for all $t \in \mathbb{R}$. It follows that the given solution actually satisfies (6.21), (6.22) for all $t \in \mathbb{R}$.

For $t < 0$ we have $q(t) < 0$ and $0 < \omega_1(t) < 1$. It follows that if one lets t run backwards, from $t = 0$ to $t = -\infty$, then one will still be in the injection (source) case, but with the mapping function $f(\zeta, t)$ non-univalent. See Sect. 7.1.2 for some further discussion of this solution.

When $0 < \omega_1 < 1$ the above polynomial solution of the Polubarinova-Galin equation does not satisfy the Löwner-Kufarev equation because the branch point $f(\omega_1(t), t)$ moves: $f(e^{3t}, t) = \frac{1}{2}e^{4t}$. However, the Löwner-Kufarev equation must also have a solution (at least a weak one). This will have a different structure, namely

$$g(\zeta, t) = b(t) \frac{(\zeta - \omega_1(t))(\zeta - \omega_2(t))(\zeta - \omega_3(t))}{(\zeta - \zeta_1(t))^2},$$

with $\zeta_1 = 1/\omega_1$. Here $0 < \omega_1 < 1$, while $\omega_{2,3} \in \mathbb{D}^e$ may be nonreal. In the notation of (2.15) and Theorem 6.1 we now have $m = 3, n = 2$ and

$$A_\infty = 0,$$

$$A_1 = 0,$$

$$A_2 = \frac{q\omega_2(1 - \omega_1\omega_2)(\omega_2 - \omega_1)}{b^2\omega_1^4(\omega_2 - \omega_3)(1 - |\omega_2|^2)(1 - \omega_2\bar{\omega}_3)},$$

$$A_3 = \frac{q\omega_3(1 - \omega_1\omega_3)(\omega_3 - \omega_1)}{b^2\omega_1^4(\omega_3 - \omega_2)(1 - |\omega_3|^2)(1 - \omega_3\bar{\omega}_2)},$$

$$A_0 = C = \frac{A_2}{\omega_2} + \frac{A_3}{\omega_3} + i\,\mathrm{Im}(\frac{A_2}{\omega_2} + \frac{A_3}{\omega_3}).$$

By insertion into (6.16), (6.17), (6.18) one gets a dynamical system for the data ω_1, ω_2, ω_3, ζ_1, b (with $\zeta_1 = 1/\omega_1$ given from outset). Despite this system looking quite complicated, the solution can, for a suitable choice of $q(t)$, be spelled out in full detail, using different tools. This will be done in Sect. 7.1.3.

Example 6.2 Let g initially be given by

$$g(\zeta, 0) = b(0)\frac{\zeta - \omega_1(0)}{\zeta - \zeta_1(0)}$$

for some $0 < \omega_1(0) < 1$, $\zeta_1(0) > 1$, $b(0) > 0$. Then, first of all, there exists a unique solution of the Polubarinova-Galin equation of the same form

$$g(\zeta, t) = b(t)\frac{\zeta - \omega_1(t)}{\zeta - \zeta_1(t)} \tag{6.23}$$

with $0 < \omega_1(t) < 1$, $\zeta_1(t) > 1$, $b(t) > 0$. The system of ordinary differential equations in Theorem 6.1 for $\omega_1(t)$, $\zeta_1(t)$, $b(t)$ explicitly becomes

$$\dot{\omega}_1 = -\frac{q\zeta_1}{b^2} - 3A_1 + \frac{2A_1\omega_1}{\omega_1 - \zeta_1},$$

$$\dot{\zeta}_1 = -\frac{q\zeta_1^2}{b^2\omega_1} - \frac{A_1\zeta_1}{\omega_1} + \frac{2A_1\zeta_1}{\omega_1 - \zeta_1},$$

$$\dot{b} = \frac{q\zeta_1}{b\omega_1} + \frac{A_1 b}{\omega_1},$$

where

$$A_1 = \frac{q(\omega_1 - \zeta_1)(1 - \omega_1\zeta_1)}{b^2(1 - |\omega_1|^2)},$$

and we have taken into account that all quantities are real. It is seen immediately
that the solution will go on as long as $\omega_1(t)$, $\zeta_1(t)$, $b(t)$ stay in the above specified
intervals. However, the so obtained solution cannot solve the Löwner-Kufarev
equation because, by Theorem 6.2, that equation requires that g has a pole of order
at least two at the reflected point of $\omega_1(t)$.

The Löwner-Kufarev equation also has a solution. This represents an evolution
on a Riemann surface $\mathscr{M}$ above $\mathbb{C}$ with a branch point over $f(\omega_1(t), t)$, which has
to remain fixed in time. If $\zeta_1(0) \neq \omega_1^*(0)$ this solution (it will be unique after the
Riemann surface $\mathscr{M}$ has been fixed) will be of the form

$$g(\zeta, t) = b(t) \frac{(\zeta - \omega_1(t))(\zeta - \omega_2(t))(\zeta - \omega_3(t))}{(\zeta - \zeta_1(t))(\zeta - \zeta_2(t))^2},$$

where $\zeta_2(t) = \omega_1(t)^*$ and $\omega_2(0) = \omega_3(0) = \zeta_2(0)$. If $\zeta_1(0) = \omega_1^*(0)$ it will be of
the slightly simpler form

$$g(\zeta, t) = b(t) \frac{(\zeta - \omega_1(t))(\zeta - \omega_2(t))}{(\zeta - \zeta_1(t))^2},$$

with $\zeta_1(t) = \omega_1(t)^*$ and $\omega_2(0) = \zeta_1(0)$. One then obtains the evolution in
Example 4.3, where ζ_1 was used as time parameter. Thus, by (4.17), (4.19), $\omega_2(t) = 2t\zeta_1(t) - \zeta_1(t)^{-1}$, $b(t) = b(0)\zeta_1(0)^{-3}\zeta_1(t)^3$.

6.3   Approach via Quadrature Identities

This approach to rational solutions has the advantage that it can incorporate
transitions of zeros through $\partial\mathbb{D}$, even when $q \neq 0$.

Connecting now to Chap. 5, we saw there that structural properties, such as
having an identity of the kind (5.2) or an estimate (5.1) holding, are preserved in
time when $f = f(\cdot, t)$ represents a weak solution. The same type of argument also
shows that the property of g being a rational function is preserved, because such
property is equivalent to an identity (5.2) holding with σ of a particularly simple
form. We start by elaborating a lemma making this statement precise.

When g is rational the computation in the beginning of the proof of Lemma 5.1
can be made more explicit and ends up with a quadrature formula for $h \in \mathscr{O}(\overline{\mathbb{D}})$.
Specifically this is

$$\frac{1}{\pi} \int_{\mathbb{D}} h|g|^2 dm = \frac{1}{2\pi i} \int_{\partial\mathbb{D}} h f^* df = \sum \operatorname*{Res}_{\mathbb{D}}(hf^*g d\zeta) + \sum_j c_j \int_{\gamma_j} hg d\zeta.$$

$$(6.24)$$

Here the γ_j are arcs in $\mathbb{D}$ connecting the points where f^* has logarithmic poles. The
above computation actually does not require h to be holomorphic in $\mathbb{D}$, it is enough
that hg is holomorphic. Thus one can allow h to have a pole at any zero of g in $\mathbb{D}$.

Equation (6.24) is a Riemann surface version, pulled back to $\mathbb{D}$, of (2.14). In the terminology of [100] the corresponding domain $\tilde{\Omega} = \tilde{f}(\mathbb{D})$ ($\tilde{f}$ being the lift of f, as in Sect. 2.2), then is a **quadrature Riemann surface**. The formula (6.24) may alternatively be presented without line integrals by expressing the right member in terms of an integral of h, namely

$$H(z) = \int_0^z h(\zeta)g(\zeta)d\zeta.$$

The quadrature identity then becomes

$$\frac{1}{\pi}\int_{\mathbb{D}} h|g|^2 dm = -\frac{1}{2\pi i}\int_{\mathbb{D}} dH \wedge d\bar{f} = -\sum \operatorname*{Res}_{\mathbb{D}}(Hdf^*) = \sum \operatorname*{Res}_{\mathbb{D}} \frac{H(\zeta)g^*(\zeta)d\zeta}{\zeta^2}.$$

By spelling out the results of the residue calculations, and taking into account the other direction we have the following.

Proposition 6.1 Let $f \in \mathcal{O}_{norm}(\overline{\mathbb{D}})$. Then $g = f'$ is a rational function if and only if there exist α_j, γ_j, a_{kj}, c_j, r, ℓ, n_j such that the quadrature identity

$$\frac{1}{\pi}\int_{\mathbb{D}} h|g|^2 dm = \sum_{j=1}^r c_j \int_{\gamma_j} hgd\zeta + \sum_{j=0}^\ell \sum_{k=1}^{n_j-1} a_{jk}h^{(k-1)}(\alpha_j) \tag{6.25}$$

holds for all $h \in \mathcal{O}(\overline{\mathbb{D}})$. Here we have used the same numbering as in (2.14). In particular, $\alpha_0 = 0$. The end points of the γ_j are the logarithmic poles of f^* and the α_j are the ordinary poles of f^*g in $\mathbb{D}$. In other words, with g of the form (2.15), these points are from the set $\{\zeta_0^*, \zeta_1^*, \ldots, \zeta_\ell^*\}$, $\zeta_0 = \infty$.

In addition, if $I \ni t \mapsto f(\cdot, t) \in \mathcal{O}_{norm}(\overline{\mathbb{D}})$ represents a weak solution of the evolution problem in Definition 4.1 and Proposition 4.2, and a quadrature identity of the kind (6.25) holds for $t = 0$, then such an identity holds for all $t \in I$, however with data depending on time: $a_{jk} = a_{jk}(t)$, $\alpha_j = \alpha_j(t)$, $\gamma_j = \gamma_j(t)$. An exception is that $\alpha_0 = 0$ is fixed, and for a_{01} we have the precise behavior $a_{01}(t) = a_{01}(0) + 2Q(t)$. (Thus $a_{01}(t)$ may become zero at one moment of time.)

Remark 6.2 The time dependence of the data, for the evolution problem, is caused by the chain rule, and disappears on using test functions of the form $h(f(\zeta, t))$, or more generally time dependent test functions $h(\zeta, t)$ satisfying (4.22) or (4.24).

The weak solution concept is based on the Löwner-Kufarev equation, but (6.25) is actually stable in time also for solutions of the Polubarinova-Galin equation. This follows from Proposition 4.1.

Proof For the first statement in the proposition, the 'only if' part follows by evaluating the residues in the previous formulas, the ζ_j being the poles of f^*g in $\mathbb{D}$. Note that a zero ω of g in $\mathbb{D}$ will allow f^* to have a pole at the same point, and of the same order, hence g to have a pole of one order higher at the reflected point

ω^*, without causing a contribution in the right member of (6.25). Alternatively, one may allow the test function h to have a pole at ω.

To prove the 'if' part we use in (6.25) the test functions

$$h(\zeta) = \frac{1}{z - \zeta} \quad (\zeta \in \mathbb{D}),$$

with $z \notin \overline{\mathbb{D}}$. Then the left member of (6.25) becomes the previously used (see (5.3)) Cauchy transform G of $|g|^2 \chi_{\mathbb{D}}$ while the right hand side takes the form $R(z) + Q(z)$, where $R(z)$ is a rational function and $Q(z)$ is the contribution from the line integrals:

$$Q(z) = \sum_j c_j \int_{\gamma_j} \frac{g(\zeta) d\zeta}{z - \zeta}.$$

Reasoning as in the proof of Lemma 5.1 we first get $G = \bar{f}g + H$ on $\overline{\mathbb{D}}$ for some $H \in \mathcal{O}(\mathbb{D})$ which is continuous up to $\partial \mathbb{D}$, and then the identity

$$\bar{f}g + H = R + Q$$

on $\partial \mathbb{D}$. The latter relation can also be written as

$$f^*(z) = \frac{R(z)}{g(z)} - \frac{H(z)}{g(z)} + \frac{Q(z)}{g(z)} \quad (z \in \partial \mathbb{D}). \tag{6.26}$$

The integrals appearing in the definition of $Q(z)$ make jumps of magnitude $\pm 2\pi i g(z)$ as z crosses γ_j from one side to the other. It follows that the first two terms in the right member of (6.26) are meromorphic functions in $\mathbb{D}$, while the last term is holomorphic except for constant $(= 2\pi i c_j)$ jumps across the arcs γ_j. These jumps disappear when differentiating (6.26). The conclusion is that $df(z) = g(z) dz$ is a rational (Abelian) differential (or f an Abelian integral), because the right member gives the appropriate extension of f to the Riemann sphere. Thus g is a rational function, as claimed. Note that (6.26) then holds identically in $\mathbb{C}$.

The second statement in the proposition, about weak solutions, is an easy consequence of (4.28).   □

Example 6.3 With

$$g(\zeta) = b \frac{(\zeta - \omega_1)(\zeta - \omega_2)}{(\zeta - \zeta_1)^2}$$

the quadrature identity is in general of the form

$$\frac{1}{\pi} \int_{\mathbb{D}} h|g|^2 dm = a_0 h(0) + a_1 h(\zeta_1^*) + c \int_0^{\zeta_1^*} hg d\zeta.$$

However, if $\zeta_1^* = \omega_1$ (or $\zeta_1^* = \omega_2$) then $a_1 = 0$ and if $\zeta_1 = \frac{1}{2}(\omega_1 + \omega_2)$ (implying that g has no residues) then $c = 0$. Both of this occurred in Example 4.3.

Taking the full Hele-Shaw evolution, as in Examples 4.2 and 4.3, into account we therefore see that one can achieve a quadrature identity description of the evolution on the unified form

$$\frac{1}{\pi} \int_{\mathbb{D}} h(\zeta)|g(\zeta, t)|^2 dm(\zeta) = 2Q(t)h(0)$$

for $0 < t < \infty$, despite the fact that $f(\zeta, t)$ changes behavior as in (4.20) when the zero of g passes through the unit circle.

Example 6.4 Similarly, with

$$g(\zeta) = b \frac{(\zeta - \omega_1)(\zeta - \omega_2)(\zeta - \omega_3)}{(\zeta - \zeta_1)^2}$$

one gets in general a quadrature identity of the form

$$\frac{1}{\pi} \int_{\mathbb{D}} h|g|^2 dm = a_{01}h(0) + a_{02}h'(0) + a_{11}h(\zeta_1^*) + c \int_0^{\zeta_1^*} hg d\zeta.$$

If g has no residues and $\zeta_1^* = \omega_1$ then the constants a_{11} and c vanish and we get just

$$\frac{1}{\pi} \int_{\mathbb{D}} h|g|^2 dm = a_{01}h(0) + a_{02}h'(0).$$

This case will be discussed in Sect. 7.1.3.

Example 6.5 We shall go one step further, this in order to connect to the example in Chap. 1, illustrated by Figs. 1.1, 1.2, and to discussions in Sect. 10.3. When

$$g(\zeta) = b \frac{(\zeta - \omega_1)(\zeta - \omega_2)(\zeta - \omega_3)(\zeta - \omega_4)}{(\zeta - \zeta_1)^2}$$

one has in general a quadrature identity of the form

$$\frac{1}{\pi} \int_{\mathbb{D}} h|g|^2 dm = a_{01}h(0) + a_{02}h'(0) + a_{03}h''(0) + a_{11}h(\zeta_1^*) + c \int_0^{\zeta_1^*} hg d\zeta.$$

If g has no residues, in other words if also its primitive f is rational, and $\zeta_1^* = \omega_1$ then, as above,

$$\frac{1}{\pi} \int_{\mathbb{D}} h|g|^2 dm = a_{01}h(0) + a_{02}h'(0) + a_{03}h''(0).$$

The map f itself will be of the form

$$f(\zeta) = \frac{b_1\zeta + b_2\zeta^2 + b_3\zeta^3 + b_4\zeta^4}{1 - \bar{\omega}_1\zeta}. \tag{6.27}$$

Compare Theorem 10.1 below.

Chapter 7
Examples

Abstract Several examples of Laplacian evolutions of a cardioid are given. One is the standard univalent evolution of the conformal map, while the others represent nonunivalent evolutions. One of these actually represents, in a certain sense, the suction case. Finally, a general discussion of injection contra suction from a Riemann surface perspective is given.

7.1 Examples: Several Evolutions of a Cardioid

In order to illustrate Theorem 5.1, as well as the structure theory in Sect. 6.2, we shall consider several different Hele-Shaw evolutions which all start out from the cardioid $\Omega(0) = f(\mathbb{D}, 0)$, where

$$f(\zeta, 0) = \zeta - \frac{1}{2}\zeta^2. \tag{7.1}$$

Thus $g(\zeta, 0) = 1 - \zeta$, having the root $\omega_1(0) = 1$ which maps onto a cusp on $\partial\Omega(0)$ at $f(1, 0) = \frac{1}{2}$. It is a major open problem to find some natural way to make Hele-Shaw suction $(q < 0)$ starting from the above cardioid, and we shall briefly discuss this problem in Sects. 7.1.4 and 7.2. We shall first construct three solutions which correspond to injection $(q > 0)$, one of them (in Sect. 7.1.3) imitating the proof of Theorem 5.1.

Coefficients of polynomials and power series of functions $f \in \mathscr{O}_{\text{norm}}(\overline{\mathbb{D}})$ will in the sequel be indexed according to

$$f(\zeta) = a_0\zeta + a_1\zeta^2 + a_2\zeta^3 + \dots \quad (a_0 > 0).$$

© The Author(s), under exclusive license to Springer Nature Switzerland AG 2021
B. Gustafsson, Y.-L. Lin, *Laplacian Growth on Branched Riemann Surfaces*,
Lecture Notes in Mathematics 2287, https://doi.org/10.1007/978-3-030-69863-8_7

7.1.1 The Univalent Solution

This is the ordinary univalent Hele-Shaw evolution $f(\cdot, t) \in \mathscr{O}_{\mathrm{univ}}(\overline{\mathbb{D}})$, which by conservation of $M_1 = a_0^2 \bar{a}_1 = -\frac{1}{2}$ is given by

$$f(\zeta, t) = a_0(t)\zeta + a_1(t)\zeta^2 = a_0(t)\zeta - \frac{1}{2a_0(t)^2}\zeta^2.$$

Adapting $q(t)$ so that $a_0(t) = e^t$, for convenience, gives

$$f(\zeta, t) = e^t \zeta - \frac{1}{2}e^{-2t}\zeta^2, \quad q(t) = e^{2t} - e^{-4t},$$

$$Q(t) = \frac{1}{4}(2e^{2t} + e^{-4t} - 3).$$

Note that $\omega_1(t) = e^{3t}$ starts out with finite speed, despite the cusp. This is possible because $q(0) = 0$. For $t > 0$, we have $q(t) > 0$.

The solution domains $\Omega(t) = f(\mathbb{D}, t)$ enjoy the quadrature identities

$$\frac{1}{\pi}\int_{\Omega(t)} h\,dm = (e^{2t} + \frac{1}{2}e^{-4t})h(0) - \frac{1}{2}h'(0) \quad (h \in \mathscr{O}(\overline{\Omega(t)})). \tag{7.2}$$

The two coefficients can be identified as $M_0(t) = 2Q(t) + \frac{3}{2}$ and M_1 respectively.

For the sake of completeness we compute the elimination function (3.30) for any polynomial function of the above kind, i.e. for any $f(\zeta) = a_0\zeta + a_1\zeta^2$ with real coefficients. A straight-forward calculation gives

$$\mathscr{E}_{f,f^*}(z, w) = \frac{a_1^4 - a_0^2 a_1^2 - a_0^2 a_1(z + w) - (a_0^2 + 2a_1^2)zw + z^2 w^2}{z^2 w^2},$$

hence the boundary of $\Omega = f(\mathbb{D})$ has equation

$$a_1^4 - a_0^2 a_1^2 - 2a_0^2 a_1 x - (a_0^2 + 2a_1^2)(x^2 + y^2) + (x^2 + y^2)^2 = 0,$$

and the Schwarz function is given by

$$S(z) = \frac{1}{2z^2}\left(a_0^2 a_1 + (a_0^2 + 2a_1^2)z \pm \sqrt{(a_0^2 a_1 + (a_0^2 + 2a_1^2)z)^2 + 4(a_1^4 - a_0^2 a_1^2)z^2}\right).$$

This applies to the solutions in this subsection and the next.

7.1.2   A Non-univalent Solution of the Polubarinova-Galin Equation

In the univalent solution, the coefficient a_0 ranges over the interval $1 \leq a_0 < \infty$, and the moment $M_0 = a_0^2 + 2|a_1|^2 = a_0^2 + \frac{1}{2}a_0^{-4}$ is an increasing function of a_0. But, as a function of a_0, M_0 is strictly convex on the entire interval $0 < a_0 < \infty$, and it has a minimum for $a_0 = 1$. Thus M_0 increases also as a_0 decreases from 1 to 0. Choosing then $a_0 = e^{-t}, 0 \leq t < \infty$, gives our second solution

$$f(\zeta, t) = e^{-t}\zeta - \frac{1}{2}e^{2t}\zeta^2, \quad q(t) = e^{4t} - e^{-2t}.$$

This is not even locally univalent, but it does solve the Polubarinova-Galin equation. The zero of $g(\zeta, t)$, $\omega_1(t) = e^{-3t}$, moves from the unit circle towards the origin, and its image point, $f(e^{-3t}, t) = \frac{1}{2}e^{-4t}$ also moves. Therefore the solution cannot be lifted to a fixed Riemann surface, and $f(\zeta, t)$ does not solve the Löwner-Kufarev equation (see Theorem 2.1).

A quadrature identity similar to (7.2) still holds. Considering first h to be defined in the image domain, it can be pulled back to $\mathbb{D}$ or be expressed in terms of the counting function (2.20). Exhibiting both we have

$$\frac{1}{\pi}\int_{\mathbb{D}} h(f(\zeta, t))|g(\zeta, t)|^2 dm(\zeta) = \frac{1}{\pi}\int_{\mathbb{C}} h\nu_{f(\cdot, t)} dm$$

$$= (e^{2t} + \frac{1}{2}e^{-4t})h(0) - \frac{1}{2}h'(0).$$

A slightly stronger form is obtained by using a time dependent test function, $h(\zeta, t)$, defined in $\mathbb{D}$ and such that (4.22) or (4.24) holds. Using (4.29) with $s = 0$ then gives

$$\frac{1}{\pi}\int_{\mathbb{D}} h(\zeta, t)|g(\zeta, t)|^2 dm(\zeta) = (e^{2t} + \frac{1}{2}e^{-4t})\,h(0, 0) - \frac{1}{2}h'(0, 0). \qquad (7.3)$$

Choosing instead h to be independent of time, the coefficient in front of h' becomes time dependent:

$$\frac{1}{\pi}\int_{\mathbb{D}} h(\zeta)|g(\zeta, t)|^2 dm(\zeta) = (e^{2t} + \frac{1}{2}e^{-4t})\,h(0) - \frac{e^t}{2}h'(0). \qquad (7.4)$$

7.1.3 A Non-univalent Solution of the Löwner-Kufarev Equation

Even though the univalent solution in Sect. 7.1.1 is perfectly good in all respects, the solution which is constructed in the proof of Theorem 5.1 is a different one, namely one which goes up on a Riemann surface with two sheets. This is because the solution in the proof is constructed in such a way that at any time, say $t = t_0$, at which a zero of $g(\zeta, t)$ reaches $\partial \mathbb{D}$, the continued solution propagates on the Riemann surface which uniformizes $f^{-1}(\zeta, t_0)$ in a neighborhood of $\overline{\mathbb{D}}$. This is a necessary step in most cases, but occasionally (as in the present example, with $t_0 = 0$) it turns out that the original Riemann surface itself is actually good enough.

Below we calculate that solution which the proof of Theorem 5.1 would have given us. This has the additional advantage of giving a reference solution which can be used as comparison in order to obtain estimates for other solutions. The idea (compare Example 6.1) is that the initial g, which we write as

$$g(\zeta, 0) = -(\zeta - 1) \cdot \frac{(\zeta - 1)(\zeta - 1)}{(\zeta - 1)^2}, \tag{7.5}$$

continues as

$$g(\zeta, t) = b(t)(\zeta - \omega_1(t)) \cdot \frac{(\zeta - \omega_2(t))(\zeta - \omega_3(t))}{(\zeta - \zeta_1(t))^2}, \tag{7.6}$$

where one of the zeros, say $\omega_1(t)$, moves into $\mathbb{D}$, $\zeta_1(t) = \omega_1^*(t)$ and where $\omega_2(t)$, $\omega_3(t)$ in addition are chosen so that $g(\zeta, t)$ has no residues. This means that $f(\zeta, t)$ will be of the form

$$f(\zeta, t) = -\frac{b_1 \zeta + b_2 \zeta^2 + b_3 \zeta^3}{\zeta - \zeta_1} \tag{7.7}$$

with $b_1 = b_1(t)$, $b_2 = b_2(t)$, $b_3 = b_3(t)$ and $\zeta_1 = \zeta(t)$, all real. The parameters b_1 and ζ_1 will turn out to be positive and strictly increasing in time. At time $t = 0$ we have

$$b(0) = -1, \tag{7.8}$$

$$b_1(0) = 1, \tag{7.9}$$

$$b_2(0) = -\frac{3}{2}, \tag{7.10}$$

$$b_3(0) = \frac{1}{2}, \tag{7.11}$$

$$\omega_1(0) = \omega_2(0) = \omega_3(0) = \zeta_1(0) = 1. \tag{7.12}$$

From (7.7) we obtain

$$g(\zeta, t) = \frac{b_1\zeta_1 + 2b_2\zeta_1\zeta + (3b_3\zeta_1 - b_2)\zeta^2 - 2b_3\zeta^3}{(\zeta - \zeta_1)^2},$$

$$f^*(\zeta, t) = \frac{b_1\zeta^2 + b_2\zeta + b_3}{\zeta^2(\zeta_1\zeta - 1)}. \tag{7.13}$$

The coefficients $b_j = b_j(t)$ and the pole $\zeta_1 = \zeta_1(t)$ are to be determined according to the following principles:

- The reflected point of $\zeta_1(t)$ is to be a zero of g:

$$g(1/\zeta_1(t), t) = 0.$$

- $f(\cdot, t)$ shall map the above point $1/\zeta_1(t)$ to a point which does not move:

$$f(1/\zeta_1(t), t) = \text{constant} = f(1, 0) = \frac{1}{2}.$$

- The moment $M_1(t)$ is conserved in time:

$$M_1(t) - \operatorname*{Res}_{\zeta=0}(ff^*g d\zeta) = M_1(0) = -\frac{1}{2}.$$

- $M_0(t)$ evolves according to

$$M_0(t) = \operatorname*{Res}_{\zeta=0}(f^*g d\zeta) = M_0(0) + 2Q(t) = \frac{3}{2} + 2Q(t).$$

The constant values $\pm\frac{1}{2}$ and $\frac{3}{2}$ above are obtained from the initial data (7.8)–(7.12). Spelling out, the above equations become

$$b_1\zeta_1^4 + 2b_2\zeta_1^3 + (3b_3\zeta_1 - b_2)\zeta_1 - 2b_3 = 0, \tag{7.14}$$

$$b_1\zeta_1^2 + b_2\zeta_1 + b_3 + \frac{1}{2}\zeta_1^2(1 - \zeta_1^2) = 0, \tag{7.15}$$

$$b_1^2 b_3 - \frac{1}{2}\zeta_1^2 = 0, \tag{7.16}$$

$$b_1 b_2\zeta_1 + 2b_2 b_3\zeta_1 + b_1 b_3(\zeta_1^2 + 2) + (\frac{3}{2} + 2Q)\zeta_1^2 = 0. \tag{7.17}$$

Here we have four equations for the five time dependent parameters b_1, b_2, b_3, ζ_1 and Q. It turns out that it is possible to solve this system by expressing all paramenters in terms of b_1:

$$\zeta_1 = +\sqrt{\frac{1}{2}(1 + 2b_1 - \frac{1}{b_1^2})},$$

$$b_2 = -\frac{\zeta_1}{4}(1 + 2b_1 + \frac{3}{b_1^2}),$$

$$b_3 = \frac{2b_1^3 + b_1^2 - 1}{4b_1^4},$$

$$Q = \frac{1}{16b_1^6}(4b_1^8 + 2b_1^7 - 12b_1^6 + b_1^4 + 6b_1^3 + 2b_1^2 - 3).$$

The range for $\zeta_1 = \zeta(t)$ is $1 \le \zeta_1 < \infty$. At time $t = 0$ we shall have $\zeta_1 = 1$. Then also $b_1 = 1$, and since one easily checks that $\frac{d\zeta_1}{db_1} > 0$ it is appropriate to fix the time scale by setting

$$b_1(t) = e^t.$$

By this all parameters b_1, b_2, b_3, ζ_1, Q become explicit functions of t. Including expansions for small $t > 0$ we have

$$\zeta_1(t) = \sqrt{\frac{1}{2}(1 + 2e^t - e^{-2t})} = 1 + t - \frac{3}{4}t^2 + \mathcal{O}(t^3), \tag{7.18}$$

$$b_1(t) = e^t = 1 + t + \frac{1}{2}t^2 + \mathcal{O}(t^3), \tag{7.19}$$

$$b_2(t) = -\frac{1}{4\sqrt{2}}(1 + 2e^t + 3e^{-2t})\sqrt{1 + 2e^t - e^{-2t}} = -\frac{3}{2} - \frac{1}{2}t + \frac{3}{8}t^2 + \mathcal{O}(t^3), \tag{7.20}$$

$$b_3(t) = \frac{1}{4}(2e^{-t} + e^{-2t} - e^{-4t}) = \frac{1}{2} - \frac{5}{4}t^2 + \mathcal{O}(t^3), \tag{7.21}$$

$$Q(t) = \frac{1}{16}(4e^{2t} + 2e^t - 12 + e^{-2t} + 6e^{-3t} + 2e^{-4t} - 3e^{-6t}) = 4t^3 + \mathcal{O}(t^4). \tag{7.22}$$

This gives

$$q(t) = \frac{1}{8}(4e^{2t} + e^t - e^{-2t} - 9e^{-3t} - 4e^{-4t} + 9e^{-6t}) = 12t^2 + \mathcal{O}(t^3),$$

$$f(\zeta, t) = -\frac{2(1 + t)\zeta - (3 + t)\zeta^2 + \zeta^3 + \mathcal{O}(t^2)}{2(\zeta - 1 - t + \mathcal{O}(t^2))}. \tag{7.23}$$

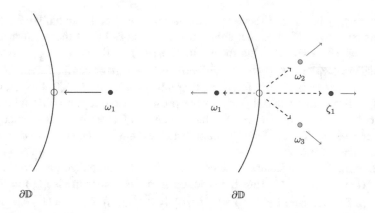

Before ω_1 reaches $\mathbb{D}$ After ω_1 has penetrated $\mathbb{D}$

Fig. 7.1 Creation of two new zeros, ω_2, ω_3, and a double pole ζ_1, as an exterior zero ω_1 penetrates into the unit disk

Thus $q(0) = \dot{q}(0) = 0$, while for $t > 0$, $q(t) > 0$, so the evolution is very slow in the beginning, in fact so slow that it is not at all singular at $t = 0$.

As for $g(\zeta, t)$, we already know (by construction) that one of its zeros is $\omega_1(t) - 1/\zeta_1(t)$. By dividing out this zero in (7.13) one gets g on the form

$$g(\zeta, t) = -2b_3(t) \frac{(\zeta - 1/\zeta_1(t))(\zeta^2 - \frac{1}{2}(b_1(t)^2 + 3)\zeta_1(t) \, \zeta + b_1(t)^3)}{(\zeta - \zeta_1(t))^2},$$

and the remaining two zeros $\omega_2(t)$, $\omega_3(t)$ are the zeros of the second degree polynomial in the numerator (Fig. 7.1). One easily checks that the discriminant of that polynomial is negative on some interval $0 < t < \varepsilon$, hence $\omega_2(t)$, $\omega_3(t)$ are non-real (a complex conjugate pair) for those values of t. For large t they are however real (the discriminant is positive). From

$$\omega_2(t) + \omega_3(t) = \frac{1}{2}(b_1(t)^2 + 3)\zeta_1(t), \quad \omega_2(t)\omega_3(t) = b_1(t)^3$$

one also realizes that the real parts of the two roots are increasing functions of t, for all $0 < t < \infty$. For small t, (7.18)–(7.22) gives the expansions

$$\omega_1(t) = 1 - t + \mathcal{O}(t^2), \tag{7.24}$$

$$\omega_{2,3}(t) = 1 + (\frac{3}{2} \pm i\frac{\sqrt{7}}{2})t + \mathcal{O}(t^2). \tag{7.25}$$

The solution $f(\zeta, t)$ represents an evolution of the cardioid which is non-univalent regarded as a map into $\mathbb{C}$ but which can be viewed as a univalent map $\tilde{f}(\zeta, t)$ into a two-sheeted Riemann surface over $\mathbb{C}$ (actually over the whole Riemann sphere $\mathbb{P}$). It is that solution which comes out of the construction in the proof of Theorem 5.1. This means that, with $f(\zeta, 0) = \zeta - \frac{1}{2}\zeta^2$, one initially considers $\mathcal{M}_0 = f(\mathbb{D}, 0) = \tilde{f}(\mathbb{D}, 0)$ as a Riemann surface over $\mathbb{C}$ and then gradually extends it, first for $\varepsilon > 0$ small, to $\mathcal{M}_\varepsilon = \tilde{f}(\mathbb{D}(0, 1 + \varepsilon), 0)$, which simply is a copy of $\mathbb{D}(0, 1 + \varepsilon)$. In the present case one can go on with this same procedure for arbitrary $\varepsilon > 0$, which eventually gives the two-sheeted covering surface $\mathcal{M} = \tilde{f}(\mathbb{P}, 0)$ of the Riemann sphere. It has branch points over $z = 1/2$ (by construction) and over $z = \infty$.

Lemma 5.3 (Conjecture) now concerns the solution pulled back to the unit disk by $f(\zeta, 0) = \zeta - \frac{1}{2}\zeta^2$. Thus the function g in that conjecture is $g(\zeta) = f'(\zeta, 0) = 1 - \zeta$. The inverse of $f(\zeta, 0) = \tilde{f}(\zeta, 0)$ is $\tilde{f}^{-1}(z, 0) = 1 - \sqrt{1 - 2z}$, hence one gets the function

$$\Phi(\zeta, t, 0) = f^{-1}(f(\zeta, t), 0) = 1 - \sqrt{1 + \frac{2}{\zeta - \zeta_1(t)}(b_1(t)\zeta + b_2(t)\zeta^2 + b_3(t)\zeta^3)},$$

which, for $0 < t < \varepsilon$ say, maps $\mathbb{D}$ conformally onto the slightly larger domain $D(t)$ (in the notation of Lemma 5.3). Because of the square root it is not entirely trivial that $\Phi(\zeta, t, 0)$ is single-valued in $\mathbb{D}$. The pole at $\zeta = \zeta_1(t)$ causes no problem in this respect since $\zeta_1(t) \notin \mathbb{D}$, but since $1 \in D(t) = \Phi(\mathbb{D}, t, 0)$ there must be a point in $\mathbb{D}$ for which the expression under the square root vanishes. However, despite this the square root does in fact resolve into a single-valued function in $\mathbb{D}$. In terms of the notations in Lemma 2.1 and Chap. 4 we have $\Phi(\zeta, t, 0) = \tilde{f}(\zeta, t)$, $f(\zeta, 0) = p(\zeta)$, $g(\zeta) = p'(\zeta)$.

The function $\Phi(\zeta, t, 0)$ is similar to the subordination functions $\varphi(\zeta, s, t)$ (for $s < t$) in (2.21), but it goes the other way. It 'superordinates' a function $f(\zeta, t)$ at a time $t > 0$ in terms of an earlier function $f(\zeta, 0)$:

$$f(\zeta, t) = f(\Phi(\zeta, t, 0), 0). \tag{7.26}$$

The inverses of the superordination functions are defined in domains $D(t) \supset \mathbb{D}$, and their restrictions to the unit disk are simply the subordination functions:

$$\varphi(\zeta, 0, t) = \Phi^{-1}(\zeta, t, 0) \quad \text{for } \zeta \in \mathbb{D}, \ t > 0.$$

The construction of $f(\zeta, t)$ is made in such a way that quadrature identities like (7.3), (7.4) remain valid. With a time dependent test function $h(\zeta, t)$, defined in $\mathbb{D}$ and satisfying (4.22) or (4.24), we have exactly the same identity (7.3) as in Sect. 7.1.2 (which is also valid for the example in Sect. 7.1.1). With h independent of time we get, as in (7.4), coefficients which depend on time, however now in a different way:

$$\frac{1}{\pi}\int_{\mathbb{D}} h(\zeta)|g(\zeta, t)|^2 dm = \left(2Q(t) + \frac{3}{2}\right)h(0) - \frac{1}{2}\frac{\sqrt{2}e^t}{\sqrt{1 + 2e^t - e^{-2t}}}h'(0).$$

This formula follows by a straightforward calculation using (7.14)–(7.17). The additional time dependent factor in the last term is simply $1/g(0, t)$ (so also in (7.4)).

We may also write the quadrature identity in a form which connects to Lemma 5.3 and the proof of Theorem 5.1: using the fixed transition function $z = f(\zeta, 0)$ between the parameter space and the image space one finds, in terms of $D(t) = \Phi(\mathbb{D}, t, 0)$,

$$\frac{1}{\pi} \int_{D(t)} h(\zeta)|g(\zeta, 0)|^2 dm = (2Q(t) + \frac{3}{2})h(0) - \frac{1}{2}h'(0). \tag{7.27}$$

This is in agreement with (4.15), and it thereby essentially confirms that the somewhat *ad hoc* attempt, starting with (7.5), for construction of a solution which uniformizes the cusp in fact gives exactly that solution which is produced in the proof of Theorem 5.1.

As for the geometry, the domains $D(t)$ are star-shaped with respect to the origin, hence the solution exists for all $0 < t < \infty$, compare [36]. Indeed, the star-shapedness follows from Eq. (5.6) in the proof of Theorem 5.1, which in the present context becomes

$$\Delta v = 4(|z - \frac{3}{4}|^2 - \frac{1}{16}), \quad z \in D(t) \setminus \mathbb{D}.$$

Here, the right member is non-negative, and as $\Delta v = 0$ in $\mathbb{D}$, v is continuous in $D(t)$ and $v = 0$ on ∂D, the required inequality $v \leq 0$ in $D(t)$ follows from the maximum principle.

7.1.4 A Solution for the Suction Case

An interesting aspect is that the now fully explicit solution $f(\zeta, t)$, defined for $0 < t < \infty$ by (7.7), (7.18), is not only smooth at $t = 0$, it even has a real analytic continuation across $t = 0$. This extended solution, defined for $-\varepsilon < t < \infty$ (say), has the drawback that it has a pole inside $\mathbb{D}$ for $t < 0$, but $q(t)$ remains positive, as can be seen from (7.23). This means that the solution represents suction out of the cardioid as t decreases to negative values.

A closer look at $f(\zeta, t)$ for $t < 0$ shows that (keeping the notation from Sect. 7.1.3) the zero $\omega_1(t)$ is now outside the unit disk, while the two complex conjugate zeros $\omega_2(t)$, $\omega_3(t)$ are inside, as well as the pole $\zeta_1(t)$. Since $f(\zeta, t)$ is no longer holomorphic in $\mathbb{D}$ we are strictly speaking outside the scope of the previously developed theory, but it is easy to see that the Polubarinova-Galin equation (2.1) still makes sense. Because of the real analyticity of all data, the fact that (2.1) holds on the interval $0 < t < \infty$ implies that it automatically holds on $-\varepsilon < t < \infty$. Of course, this can also be verified by a direct (but quite tedious) calculation. The

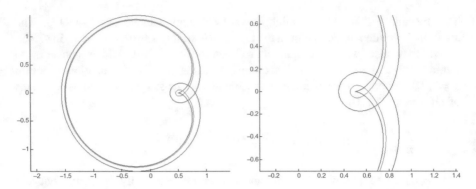

Fig. 7.2 Solution described in Sect. 7.1.3 (normal size and enlarged). Numerics and graphics are produced by Joakim Roos

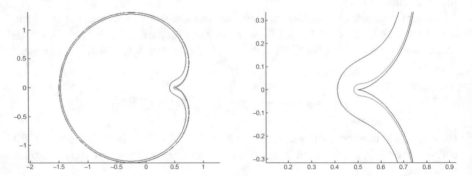

Fig. 7.3 Solution described in Sect. 7.1.4 (normal size and enlarged). Numerics and graphics are produced by Joakim Roos

boundary curve $f(\partial\mathbb{D}, t)$ is for each $t < 0$ a smooth closed loop contained in $\Omega(0)$, and it recedes as t decreases (Figs. 7.2, 7.3).

To get a clear picture of the situation, note first that, for $t \neq 0$, $f(\zeta, t)$ is a rational function of order three, hence it maps the Riemann sphere $\mathbb{P}$ onto the sphere covered thrice. Viewed as a covering map, f has branch points over $\beta_j = f(\omega_j(t), t)$, $j = 1, 2, 3$, and $\beta_4 = \infty$. Obviously, β_4 does not depend on t, and the same is actually true also for β_1, by the construction of f in Sect. 7.1.3. In fact, $\beta = 1/2 =$ the cusp point of $\partial\Omega(0)$. As in previous situations we shall write $\tilde{f}$ when we think of f as a univalent map into a Riemann surface.

To make it perfectly clear, we first have a trivial decomposition of f as

$$\mathbb{P} \xrightarrow{\mathrm{id}} \mathbb{P} \xrightarrow{f(\cdot, t)} \mathbb{P}.$$

Then the idea is to consider the middle $\mathbb{P}$ as an abstract Riemann surface, denoted $\mathscr{F}(t)$, and think of the last map as a covering projection. The first map will then be called $\tilde{f}(\cdot, t)$:

$$\mathbb{P} \xrightarrow{\tilde{f}(\cdot, t)} \mathscr{F}(t) \xrightarrow{\text{proj}} \mathbb{P}.$$

Because of the covering projection, $\mathscr{F}(t)$ is more than an abstract Riemann surface, it inherits a Riemannian metric from $\mathbb{P}$. And as a Riemannian manifold it really depends on t. Using the variable ζ in $\tilde{z} = \tilde{f}(\zeta, t)$ (where $\tilde{z} \in \mathscr{F}(t)$) as a coordinate, the metric on $\mathscr{F}(t)$ is given by

$$ds^2 = |f'(\zeta, t)|^2 |d\zeta|^2. \tag{7.28}$$

The branch points, together with appropriate 'cuts' between them, define an exact division of $\mathscr{F}(t)$ into sheets $\mathbb{P}^{(j)}$ (copies of $\mathbb{P}$) such that the pre-image $\tilde{f}^{-1}(a, t)$ of any point $a \in \mathbb{P}$ has one point on each sheet. If a is a branch point, two or more of these pre-images are common to some sheets. We may write the above as

$$\mathscr{F}(t) = (\mathbb{P}^{(1)} \cup \mathbb{P}^{(2)} \cup \mathbb{P}^{(3)})/\{\frac{1}{2}, \beta_2(t), \beta_3(t), \infty\}. \tag{7.29}$$

We have assumed that $t \neq 0$, and we choose the numbering so that $\mathbb{P}^{(1)}$ and $\mathbb{P}^{(2)}$ are connected by the branch points at $\beta_1 = 1/2$ and ∞, with the major part of $\tilde{\Omega} = \tilde{f}(\mathbb{D}, t)$ lying in $\mathbb{P}^{(1)}$. The third branch $\mathbb{P}^{(3)}$ is then connected to the previous ones via the branch points $\beta_2(t)$ and $\beta_3(t)$. To be precise about the cuts one may, for example, have one cut along the positive real axis, from $1/2$ to ∞, which serves as a passage between $\mathbb{P}^{(1)}$ and $\mathbb{P}^{(2)}$, and another cut between $\beta_2(t)$ and $\beta_3(t)$, serving as a passage to $\mathbb{P}^{(3)}$. The pre-images of these cuts under $f^{-1}(\cdot, t)$ are curves which divide $\mathbb{P}$ into three pieces, thereby they also divide $\mathscr{F}(t)$ into three pieces, which then become $\mathbb{P}^{(1)}, \mathbb{P}^{(2)}, \mathbb{P}^{(3)}$. See Fig. 7.4.

When $t = 0$ there are only two copies of $\mathbb{P}$, say

$$\mathscr{M} = \tilde{f}(\mathbb{P}, 0) = (\mathbb{P}^{(1)} \cup \mathbb{P}^{(2)})/\{\frac{1}{2}, \infty\},$$

since $f(\zeta, 0) = \zeta - \frac{1}{2}\zeta^2$ has order two. The notation $\mathscr{M}$ for this surface was introduced in Sect. 7.1.3. Thus the geometric meaning of changing $g(\zeta, 0) = 1 - \zeta$ to the form (7.5) is that one adds a new Riemann sphere $\mathbb{P}^{(3)}$, which for $t = 0$ is disconnected from $\mathbb{P}^{(1)}$ and $\mathbb{P}^{(2)}$, but which gets attached via the branch points $\beta_2(t), \beta_3(t)$ when t moves away from $t = 0$. For $t = 0$ it is convenient to define $\mathscr{F}(0)$ to be the disconnected Riemann surface corresponding to (7.5):

$$\mathscr{F}(0) = \mathscr{M} \cup \mathbb{P}^{(3)}.$$

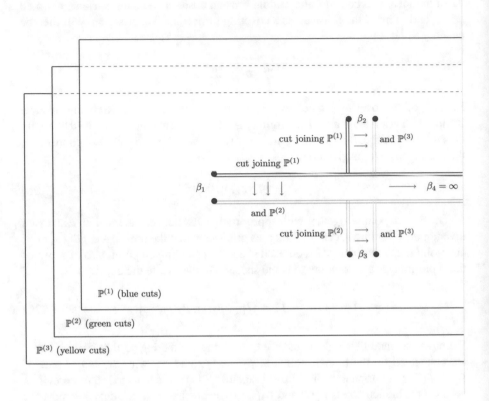

Fig. 7.4 Schematic picture of branch points and possible choices of branch cuts for Sect. 7.1.4

For $0 < t < \varepsilon$, $\tilde{\Omega}(t) = \tilde{f}(\mathbb{D}, t)$ lies mainly in $\mathbb{P}^{(1)}$ (choosing the numbering this way), but also goes up with a small piece, around $\beta_1 = 1/2$, to $\mathbb{P}^{(2)}$. For example, for the counting function (2.20) we have $\nu_{f(\cdot,t)} = 2$ in a certain neighborhood of $z = 1/2$, while $\nu_{f(\cdot,t)} = 1$ in the remaining part of $\Omega(t) = f(\mathbb{D}, t)$.

Even though the Riemannian manifold $\mathscr{F}(t)$ as a whole depends on t, there is, when $0 < t < \varepsilon$, a certain neighborhood, say $\mathscr{M}_\varepsilon$, of the closure of $\tilde{\Omega}(t) = \tilde{f}(\mathbb{D}, t)$ which does not. This is because the branch points $\beta_2(t)$, $\beta_3(t)$, which are responsible for the time dependence, are outside $\mathscr{M}_\varepsilon$. A more precise argument for the time independence can be given in terms of superordination: instead of using, as in (7.28), ζ as a local variable we can in $\mathscr{M}_\varepsilon$ use $z = \Phi(\zeta, t, 0)$. This gives, by (7.26),

$$f'(\zeta, t)d\zeta = f'(\Phi(\zeta, t, 0), 0)\Phi'(\zeta, t, 0)d\zeta = f'(z, 0)dz,$$

which brings the metric in $\mathscr{M}_\varepsilon$ on the time independent form

$$ds^2 = |f'(z, 0)|^2 |dz|^2.$$

When $-\varepsilon < t < 0$, on the other hand, $\omega_2(t)$, $\omega_3(t) \in \mathbb{D}$ while $\omega_1(t) \in \mathbb{D}^e$, and it can be seen (by an argument to be given in Sect. 7.2) that

$$\tilde{\Omega}(t) = (\Omega^{(1)}(t) \cup \mathbb{P}^{(3)})/\{\beta_2(t), \beta_3(t)\}, \tag{7.30}$$

where $\Omega^{(1)}(t) \subset \mathbb{P}^{(1)}$ is a simply connected subdomain of $\Omega(0)$ which connects to $\mathbb{P}^{(3)}$ via $\beta_2(t), \beta_3(t) \in \Omega^{(1)}(t)$. Thus

$$\tilde{\Omega}^e(t) = ((\mathbb{P}^{(1)} \setminus \overline{\Omega^{(1)}(t)}) \cup \mathbb{P}^{(2)})/\{\frac{1}{2}, \infty\},$$

and since the branch points here are fixed there is a neighborhood, say $\mathcal{N}_\varepsilon$, of $\tilde{\Omega}^e(t)$ in $\mathscr{F}(t)$ on which the short time evolution takes place and can be generated by partial balayage.

A consequence of the above is that, when $-\varepsilon < t < 0$, $\partial\tilde{\Omega}(t) = \partial\Omega^{(1)}(t)$ is a simple closed curve which lies entirely in $\Omega(0)$. Forgetting about $\mathbb{P}^{(3)}$ in (7.30), it is tempting to view $\Omega^{(1)}(t)$ as a result of suction from $\Omega(0)$. However, since the surrounding Riemann surface is time dependent, as one sees from (7.30), this will be only a rather relaxed form of suction, satisfying the Polubarinova-Galin, but not the Löwner-Kufarev, equation. From a classical perspective it is always impossible to suck from a domain having that type of cusp $\Omega(0)$ has (compare [103]), but allowing extra sheets and movable branch points makes the situation more flexible. We shall discuss these matters in some generality in the next section.

When projected to the Riemann sphere, the above boundaries $\partial\tilde{\Omega}(t)$ are algebraic curves, both for positive and negative values of t. The polynomial equations for the curves can in principle be found by calculating the elimination function, like in Sect. 7.1.1, but since this would involve solving a third order algebraic equation we have refrained from any attempts in that direction.

7.2   Injection Versus Suction in a Riemann Surface Setting

Recall the meaning of the Polubarinova-Galin equation (2.1), or (4.7), in terms of the Green function for the Riemann surface domain $\tilde{\Omega}(t) = \tilde{f}(\mathbb{D}, t)$, as spelled out in (4.9):

$$\dot{\tilde{f}}_{\text{normal}}(\tilde{z}, t) = q(t)|\nabla G_{\tilde{\Omega}(t)}(\tilde{z}, \tilde{0})|, \quad \tilde{z} \in \partial\tilde{\Omega}(t). \tag{7.31}$$

The Green function is simply obtained from that of the unit disk, by conformal invariance:

$$G_{\tilde{\Omega}(t)}(\tilde{z}, \tilde{0}) = -\log|\zeta|, \tag{7.32}$$

where $\tilde{z} = \tilde{f}(\zeta, t)$.

Now assume that f is a rational function, of order r say. So we assume that $g = f'$ is a rational function free of residues. In terms of the structure (2.15) this means that the number m there lies in the interval $r - 1 \leq m \leq 2(r - 1)$, where the extreme cases correspond to f having a single pole of order r (then f must be polynomial of degree r), respectively f having r distinct simple poles. When we consider f as a conformal map into a Riemann surface, and denote it by $\tilde{f}$ instead, then it maps $\mathbb{P}$ onto an r-fold covering surface $\mathscr{F}(t)$ of $\mathbb{P}$. As explained in the previous section, after a choice of cuts we can think of $\mathscr{F}(t)$ as consisting of r copies of $\mathbb{P}$ connected by branch points:

$$\tilde{f}(\cdot, t) : \mathbb{P} \to \mathscr{F}(t) = (\mathbb{P}^{(1)} \cup \cdots \cup \mathbb{P}^{(r)})/\{\text{branch points}\}.$$

Now $\tilde{\Omega}(t) = \tilde{f}(\mathbb{D}, t) \subset \mathscr{F}(t)$. Let $\tilde{\Omega}^e(t) = \tilde{f}(\mathbb{D}^e, t)$ be the complementary domain in $\mathscr{F}(t)$. Since the Green function for $\mathbb{D}^e$ with pole at infinity is $\log |\zeta|$ it follows that

$$G_{\tilde{\Omega}^e(t)}(\tilde{z}, \tilde{\infty}) = \log |\zeta|,$$

with $\tilde{z} = \tilde{f}(\zeta, t)$ as before, and $\tilde{\infty} = \tilde{f}(\infty, t) \in \mathscr{F}(t)$. Thus, comparing with (7.32), $G_{\tilde{\Omega}^e}(\tilde{z}, \tilde{\infty})$ is after a sign change simply the harmonic continuation of $G_{\tilde{\Omega}}(\tilde{z}, \tilde{0})$. In particular it follows that, on the common boundary $\partial \tilde{\Omega}(t) = \partial \tilde{\Omega}^e(t)$,

$$|\nabla G_{\tilde{\Omega}^e}(\tilde{z}, \tilde{\infty})| = |\nabla G_{\tilde{\Omega}}(\tilde{z}, \tilde{0})|.$$

Returning to (7.31) this means that we also have

$$\dot{\tilde{f}}_{\text{normal}}(\tilde{z}, t) = q(t)|\nabla G_{\tilde{\Omega}^e(t)}(\tilde{z}, \tilde{\infty})|, \quad \tilde{z} \in \partial \tilde{\Omega}^e(t). \tag{7.33}$$

The left member here is the same as in (7.31), but in relation to $\tilde{\Omega}^e(t)$ it is an inward pointing normal vector.

Thus the equation which describes injection at $\tilde{0}$ in $\tilde{\Omega}(t)$ at the same time describes suction at $\tilde{\infty}$ from $\tilde{\Omega}^e(t)$, and conversely. So the two problems are in the present setting equivalent, which might seem remarkable since the suction problem is known in general to be highly unstable and ill-posed, while injection always is stable and well-posed. The explanation is that we have lifted everything to a Riemann covering surface $\mathscr{F}(t)$, which has branch points allowed to move, and which is conformally equivalent to the Riemann sphere in such a way that $\tilde{\Omega}(t)$ and $\tilde{\Omega}^e(t)$ correspond to $\mathbb{D}$ and $\mathbb{D}^e$, respectively. With movable branch points in $\tilde{\Omega}(t)$ the solution of (7.31) is not unique, as is clear from Theorem 2.2. Similarly for $\tilde{\Omega}^e(t)$ and (7.33). And some branch points must be allowed to move because $\mathscr{F}(t)$ as a whole is time dependent.

Thus lifting to a Riemann surface with movable branch points can be viewed as a kind of relaxation, opening up for more suction solutions. Indeed, if we can arrange that the branch points in $\tilde{\Omega}^e(t)$ are fixed, then we may perform injection at

$\tilde{\infty} = \tilde{f}(\infty, t) \in \tilde{\Omega}^e(t)$ by partial balayage, and this will correspond, in some sense, to suction at $\tilde{0}$ from $\tilde{\Omega}(t)$.

As an example, we can explain the suction from $\Omega(0)$ obtained at the end of Sect. 7.1.4. We identify this initial domain, which has a cusp at $z = 1/2$, with $\tilde{\Omega}(0) \subset \mathscr{F}(0)$, which lies entirely in $\mathbb{P}^{(1)}$, in the notation of Sect. 7.1.4. The complementary domain in $\mathscr{F}(0)$ then is $\tilde{\Omega}^e(0) = ((\mathbb{P}^{(1)} \setminus \overline{\tilde{\Omega}(0)}) \cup \mathbb{P}^{(2)})/\{1/2, \infty\} \cup \mathbb{P}^{(3)}$.

Now, each of $\mathbb{P}^{(1)}$, $\mathbb{P}^{(2)}$ and $\mathbb{P}^{(3)}$ has its own point of infinity, denote them by $\tilde{\infty}^{(1)}$, $\tilde{\infty}^{(2)}$, $\tilde{\infty}^{(3)}$, respectively, and we must choose from which of these to inject. But there is actually no choice, it is impossible to inject at $\tilde{\infty}^{(3)}$ because $\mathbb{P}^{(3)}$ is isolated (at this initial stage $t = 0$), and the other two points are actually the same point in $\mathscr{F}(0)$ since they represent the branch point β_4 for $f(\zeta, 0)$. In fact, $\tilde{\infty}^{(1)} = \tilde{\infty}^{(2)} = \tilde{f}(\infty, 0)$, which is the correct source point.

In terms of partial balayage on $\mathscr{M} = \mathscr{F}(0)$, the desired evolution $\tilde{\Omega}^e(t)$, with initial domain $\tilde{\Omega}^e(0)$ and source at $\tilde{\infty}^{(1)} = \tilde{\infty}^{(2)} = \tilde{f}(\infty, 0)$, in principle becomes

$$\mathrm{Bal}(2\pi Q(-t)\delta_{\tilde{\infty}^{(1)}} + \chi_{\tilde{\Omega}^e(0)}\tilde{m}, \tilde{m}) = \chi_{\tilde{\Omega}^e(t)}\tilde{m}. \tag{7.34}$$

Here we are using the same time variable as in Sect. 7.1.4, which means that $-\varepsilon < t < 0$ and $Q(-t) > 0$.

Unfortunately, (7.34) does not really make sense because the measure we are sweeping has infinite mass. But it is easy to remedy the situation by subtracting $\chi_{\tilde{\Omega}^e(0)}\tilde{m}$ from both sides. Using that $\partial\tilde{\Omega}(0)$ is a nullset with respect to $\tilde{m}$ this gives

$$\mathrm{Bal}(2\pi Q(-t)\delta_{\tilde{\infty}^{(1)}}, \chi_{\tilde{\Omega}(0)}\tilde{m}) = \chi_{\tilde{\Omega}(0) \setminus \tilde{\Omega}(t)}\tilde{m}, \tag{7.35}$$

where then $\tilde{\Omega}(t)$ is the result of the suction out of $\tilde{\Omega}(0)$. Equation (7.35) makes perfectly good sense for $-\varepsilon < t < 0$, where $\varepsilon > 0$ is chosen so that $2\pi Q(\varepsilon) = \tilde{m}(\tilde{\Omega}(0))$, and it is the correct formula for the describing the evolution of $\tilde{\Omega}(t)$ (or $\tilde{\Omega}^e(t)$). Of course, everything can be pulled back from $\mathscr{M}$ to $\mathbb{P}$ by using the variable ζ in $\tilde{z} = \tilde{f}(\zeta, 0)$ as coordinate on $\mathscr{M}$. Then one gets

$$\mathrm{Bal}(2\pi Q(-t)\delta_\infty, |g|^2\chi_\mathbb{D}) = |g|^2\chi_{\mathbb{D}\setminus\hat{D}(t)} \qquad (-\varepsilon < t < 0),$$

where $g(\zeta) = g(\zeta, 0) = 1 - \zeta$ and $\hat{D}(t) = \tilde{f}^{-1}(\tilde{\Omega}(t), 0)$.

Chapter 8
Moment Coordinates and the String Equation

Abstract The Polubarinova-Galin is identified with what is called the string equation in contexts of integrable hierarchies. In the non-univalent case the string equation requires further specification. We discuss the arising difficulties in this respect and illustrate by some examples.

8.1 The Polubarinova-Galin Equation as a String Equation

In the remaining chapters we shall discuss the Polubarinova-Galin equations and corresponding Laplacian growth from slightly different points of view. More precisely we shall embed Laplacian growth in a larger class of domain variations arising naturally when one uses the harmonic moments as parameters for simply connected domains. This turns out to be an interesting change of perspective, leading into areas of mathematical physics such as theories of integrable systems and conformal field theory. Our treatment is based on developments starting in the 1990s by M. Mineev-Weinstein, P. Wiegmann, A. Zabrodin and others. A few early references are [61, 64, 74, 79, 124].

The Polubarinova-Galin equation reappears now from pure mathematical consideration, without any viscous fluid present and even without any time variable, under the name **string equation**. Written in terms of the appropriate Poisson bracket, which is a variant of (2.9), it is

$$\{f, f^*\} = 1, \tag{8.1}$$

to be interpreted as an identity for normalized conformal maps f from the unit disk. This is to be compared with the Polubarinova-Galin equation on the form (2.7).

The name string equation can be traced back to theories of integrable hierarchies, such as the 2D Toda hierarchy, for which there appears a pair $L, \bar{L}$ of "Lax operators" satisfying what in that context is called the string equation, namely

$$[L, \bar{L}] = \hbar,$$

© The Author(s), under exclusive license to Springer Nature Switzerland AG 2021
B. Gustafsson, Y.-L. Lin, *Laplacian Growth on Branched Riemann Surfaces*,
Lecture Notes in Mathematics 2287, https://doi.org/10.1007/978-3-030-69863-8_8

where $\hbar > 0$ is Planck's constant. As $\hbar \to 0$ this equation is taken over, in situations described in [74, 124] (for example), by our "dispersionless" string equation (8.1) for a conformal map. In this way the subject of Laplacian growth connects to topological gravity and matrix models of 2D gravity. See papers mentioned above, and in addition [58, 75, 76]. Somewhat related are also connections to ensembles of random normal matrix, quantum Hall regimes and Coulomb gas ensembles, see [50, 114, 126, 127].

The main question for the present treatment is what happens if f is no longer univalent. Does the string equation still make sense and, if so, does it hold? As we shall see, the answer is that the Poisson bracket does not automatically make sense in the non-univalent case, but that one can extend its meaning, actually in several different ways, and after such a step the string equation indeed holds. These matters are closely related to the discussions of the Polubarinova-Galin equation versus the Löwner-Kufarev equation of in Sect. 2.3. We shall show (in Chap. 10) that the string equation makes sense and holds for a class of rational functions related to quadrature Riemann surfaces (in the spirit of Sect. 6.3), and also that it holds, in a different sense, for polynomials. Our treatment closely follows [33, 34].

8.2   The String Equation for Univalent Conformal Maps

We consider analytic functions $f(\zeta)$ defined in a neighborhood of the closed unit disk and with the usual normalization, $f \in \mathcal{O}_{\mathrm{norm}}(\overline{\mathbb{D}})$. We start by looking at the univalent case, i.e. $f \in \mathcal{O}_{\mathrm{univ}}(\overline{\mathbb{D}})$, with f mapping $\mathbb{D}$ onto a single-sheeted domain $\Omega = f(\mathbb{D})$. Like in Chap. 7 we write the Taylor expansions around the origin on the form

$$f(\zeta) = a_0\zeta + a_1\zeta^2 + a_2\zeta^3 + \dots \quad (a_0 > 0).$$

Recall from (2.4) the harmonic moments of Ω:

$$M_k = \frac{1}{\pi} \int_{\Omega} z^k dm(z) = \frac{1}{2\pi i} \int_{\partial \mathbb{D}} f(\zeta)^k f^*(\zeta) f'(\zeta) d\zeta, \qquad (8.2)$$

where, as usual,

$$f^*(\zeta) = \overline{f(1/\bar{\zeta})} \qquad (8.3)$$

denotes the holomorphic reflection in the unit circle. In the form of the rightmost member in (8.2) the moments make sense also when f is not univalent. In addition, they make sense for negative values of k, in which case they can be interpreted as moments of the complementary domain, namely $\Omega^e = \mathbb{P} \setminus \overline{\Omega}$ (in case f is univalent).

Computing the last integral in (8.2) by residues gives **Richardson's formula** [88] for the moments:

$$M_k = \sum_{(j_0,\dots,j_k) \geq (0,\dots,0)} (j_0 + 1) a_{j_0} \cdots a_{j_k} \bar{a}_{j_0+\dots+j_k+k}. \tag{8.4}$$

This is a highly nonlinear relationship between the coefficients of f and the moments, and even if f is a polynomial of low degree it is virtually impossible to invert it, to obtain $a_k = a_k(M_0, M_1, \dots)$, as would be desirable in many situations.

Example 8.1 Even in the simplest nontrivial case, with a second degree polynomial $f(\zeta) = a_0\zeta + a_1\zeta^2$, where

$$\begin{cases} M_0 = a_0^2 + 2a_1\bar{a}_1, \\ M_1 = a_0^2\bar{a}_1, \end{cases}$$

it is difficult to write a_0, a_1 explicitly in terms of M_0, M_1. Indeed, this would require solving a third degree polynomial equation, which we refrain from doing. As a further issue, a_0, a_1 will not be holomorphic with respect to M_1, so one actually gets

$$\begin{cases} a_0 = a_0(M_0, M_1, \bar{M}_1), \\ a_1 = a_1(M_0, M_1, \bar{M}_1). \end{cases}$$

Thus $f(\zeta) = f(\zeta, M_0, M_1, \bar{M}_1)$, to exhibit analytic dependence. In algebraic or analytic relations like the above, M_1 and $\bar{M}_1$ can be treated as independent variables. The same for a_j and $\bar{a}_j$ when $j \geq 1$, so one should actually add $\bar{a}_1 = \bar{a}_1(M_0, M_1, \bar{M}_1)$ to the above two equations.

Despite what has been said above there is, quite remarkably, an explicit expression for the Jacobi determinant of the change $(a_0, a_1, \dots) \mapsto (M_0, M_1, \dots)$ when f is restricted to the class of polynomials of a fixed degree. This formula, which was proved by O. Kuznetsova and V. Tkachev [66, 116] after an initial conjecture of C. Ullemar [117], will be discussed in some depth in Chap. 10.

There are examples of different solid domains (see before Theorem 3.3 for the terminology) having the same harmonic moments, see examples in [94, 99, 128], but it is also known that for domains having analytic boundary the harmonic moments are sensitive for at least small variations of the domain. Specific arguments were given in Chap. 3, namely in Theorem 3.2 (independence of moments) and Theorem 3.3 (completeness of moments). And for polynomial maps f the above mentioned Jacobi determinant is indeed nonzero, as we shall see in Chap. 10.

It follows that the harmonic moments can be viewed as local coordinates for the family of simply connected domains with analytic boundary, or for the corresponding space of univalent functions. Hence we can write

$$f(\zeta) = f(\zeta, M_0, M_1, M_2, \dots),$$

or more precisely $f(\zeta) = f(\zeta, M_0, M_1, \bar{M}_1, M_2, \bar{M}_2, \dots)$, if one wants to exhibit analytic dependence. In particular, derivatives like $\partial f / \partial M_k$ and $\partial f / \partial \bar{M}_k$ make sense.

Now we are in a position to define the Poisson bracket. In this definition the functions involved need only be defined in a neighborhood of the unit circle, but in most applications each function will be analytic either in the (closed) disk or in the (closed) exterior disk (including the point of infinity).

Definition 8.1 For any two functions, $f_1(\zeta) = f_1(\zeta, M_0, M_1, M_2, \dots)$ and $f_2(\zeta) = f_2(\zeta, M_0, M_1, M_2, \dots)$, which are analytic in a neighborhood of the unit circle and which are parametrized by the moments, we define

$$\{f_1, f_2\} = \zeta \frac{\partial f_1}{\partial \zeta} \frac{\partial f_2}{\partial M_0} - \zeta \frac{\partial f_2}{\partial \zeta} \frac{\partial f_1}{\partial M_0}. \tag{8.5}$$

This is again a function analytic in a neighborhood of the unit circle and parametrized by the moments.

We have previously introduced a similar Poisson bracket indexed by a time variable t for any given evolution, see (2.9), and in terms of that the Polubarinova-Galin equation (2.1) takes the form (2.7), i.e.

$$\{f, f^*\}_t = 2q(t).$$

Solutions of this equation preserve the harmonic moments M_k for $k \geq 1$, while $M_0(t)$ changes with speed $2q(t)$, see (2.2) or (2.5). Hence it follows, somewhat indirectly, that the string equation (8.1) indeed holds for normalized univalent functions $f \in \mathcal{O}_{\text{univ}}(\overline{\mathbb{D}})$. More direct proofs of the string equation (in the univalent case) are given in [32, 43], and in [124] for the opposite geometry.

8.3   Intuition and Physical Interpretation in the Non-univalent Case

The main issue for the present treatment is that we wish to allow non-univalent analytic functions in the string equation. Then there will arise the problem that f is not determined by the moments $M_0, M_1, \dots$ alone. Since $\partial f / \partial M_0$ is a partial derivative one has to specify all other independent variables in order to give a meaning to it. So there may be more variables, say

$$f(\zeta) = f(\zeta, M_0, M_1, \dots, B_1, B_2, \dots).$$

It turns out that the string equation still holds, independent of how $B_1, B_2, \ldots$ are chosen, but the meaning of the string equation then depends on the choice of these extra variables. This should actually not come as a surprise since the string equation is equivalent to the Polubarinova-Galin equation and we have seen in Sect. 2.3 that this equation does not determine the evolution completely.

What does determine the evolution completely is the Löwner-Kufarev equation (2.10), (2.11), see Theorem 2.1. In view of that theorem a natural choice for the extra variables should be the locations of branch points, i.e. one takes $B_j = f(\omega_j)$, where the $\omega_j \in \mathbb{D}$ denote zeros of $f'(\zeta)$ inside $\mathbb{D}$. Keeping these branch points fixed, as is implicit then in the notation $\partial/\partial M_0$, means that f can be viewed as a conformal map into a fixed Riemann surface, which will be a branched covering over the complex plane.

There are also other possibilities for giving a meaning to the string equation, like restricting f to the class of polynomials of a fixed degree. But then one must allow the branch points to move, so that gives a different meaning to $\partial/\partial M_0$. There are actually more problems in the non-univalent case. Even if we specify all branch points, the test class of functions $1, z, z^2, \ldots$ used in defining the moments may be too small since each of these functions takes the same value on all sheets above any given point in the complex plane. In order for the Riemann surface and the conformal map to be determined one would need all analytic functions on the Riemann surface itself as test functions.

For polynomial maps f there are only finitely many nonzero moments. The same occurs for certain kinds of (non-univalent) rational functions, and for the rigorous treatment in Chap. 10 we shall concentrate on such cases, where only finitely many moments are nonzero.

Like in previous chapters we consider non-univalent analytic functions as conformal maps into Riemann surfaces above $\mathbb{C}$, in general with branch points, and the non-univalence is then absorbed in the covering projection. It is easy to understand from general considerations that such a Riemann surface, or the corresponding conformal map, will in general not be determined by the moments $M_0, M_1, M_2, \ldots$ alone.

As a simple example, consider an oriented curve Γ in the complex plane encircling a certain point, a, twice. In terms of the winding number, or index,

$$\nu_\Gamma(z) = \frac{1}{2\pi i} \int_\Gamma d\log(\zeta - z) \quad (z \in \mathbb{C} \setminus \Gamma), \tag{8.6}$$

this means that $\nu_\Gamma(a) = 2$. If $\Gamma = f(\partial\mathbb{D})$, $f \in \mathcal{O}_{\text{norm}}(\overline{\mathbb{D}})$, the winding number is the same as the previously defined counting function for f, see Definition 2.1.

Having only the curve Γ available it is natural to define the harmonic moments for the multiply covered (with multiplicities ν_Γ) set inside Γ as

$$M_k = \frac{1}{\pi} \int_\mathbb{C} z^k \nu_\Gamma(z) dm = \frac{1}{2\pi i} \int_\Gamma z^k \bar{z} dz, \quad k = 0, 1, 2, \ldots.$$

This agrees with (8.2) if $\Gamma = f(\partial\mathbb{D})$. It is tempting to think of this integer weighted set as a Riemann surface over (part of) the complex plane. However, without further information this is not possible. Indeed, since some points have index ≥ 2 such a covering surface will have to have branch points, and these have to be specified in order to make the set into a Riemann surface. And only after that we have a uniquely determined conformal map f. Thus f is in general not determined by the moments alone. In the simplest non-univalent cases f will be (locally) determined by the harmonic moments together with the location of the branch points.

The above discussion may be illustrated by Fig. 1.2 with a being the branch point. If one moves a, without changing the boundary curve, the conformal map to the multi-sheeted domain inside the curve will change. In the next section we shall elaborate the above ideas in detail in a specific example.

8.4   An Example

8.4.1   General Case

We shall consider again the Examples 4.2, 4.3 in Chap. 4, but now in somewhat more generality. For constants $a, b, c \in \mathbb{C}$ with $0 < |a| < 1 < |b|$, $c \neq 0$, consider the rational function

$$f(\zeta) = c \cdot \frac{\zeta(\zeta - a)}{\zeta - b}. \tag{8.7}$$

Here the derivative

$$f'(\zeta) = c \cdot \frac{\zeta^2 - 2b\zeta + ab}{(\zeta - b)^2} = c \cdot \frac{(\zeta - \omega_1)(\zeta - \omega_2)}{(\zeta - b)^2}$$

vanishes for

$$\omega_{1,2} = b(1 \pm \sqrt{1 - \frac{a}{b}}), \tag{8.8}$$

where $\omega_1\omega_2 = ab$, $\frac{1}{2}(\omega_1 + \omega_2) = b$. The constant c is to be adapted according to the normalization

$$f'(0) = \frac{ac}{b} > 0.$$

This fixes the argument of c, so the parameters a, b, c represent 5 real degrees of freedom for f.

We will be interested in choices of a, b, c for which one of the roots $\omega_{1,2}$ is in the unit disk, say $|\omega_1| < 1$. Then $|\omega_2| > 1$. The function f is in that case not locally univalent, but can be considered as a conformal map into a Riemann surface over $\mathbb{C}$ having a branch point over

$$B_1 = f(\omega_1) = cb\left(1 - \sqrt{1 - \frac{a}{b}}\right)^2 = \frac{c\,\omega_1^2}{b}. \tag{8.9}$$

Here we let, as a matter of notation, ω_1 correspond to the minus sign in (8.8) (this is natural in the case $0 < a < 1 < b$). The holomorphically reflected function is

$$f^*(\zeta) = \bar{c} \cdot \frac{1 - \bar{a}\zeta}{\zeta(1 - \bar{b}\zeta)}.$$

Let $H(z)$ be any analytic (test) function defined in a neighborhood of the closure of $f(\mathbb{D})$, for example $H(z) = z^k$, $k \geq 0$. Then, denoting by v_f the index of $f(\partial\mathbb{D})$, see (8.6), we have

$$\frac{1}{\pi} \int_{\mathbb{C}} H v_f dm = \frac{1}{2\pi i} \int_{\mathbb{D}} H(f(\zeta))|f'(\zeta)|^2 d\bar{\zeta} d\zeta = \frac{1}{2\pi i} \int_{\partial\mathbb{D}} H(f(\zeta)) f^*(\zeta) f'(\zeta) d\zeta$$

$$= \operatorname*{Res}_{\zeta=0} H(f(\zeta)) f^*(\zeta) f'(\zeta) d\zeta + \operatorname*{Res}_{\zeta=1/\bar{b}} H(f(\zeta)) f^*(\zeta) f'(\zeta) d\zeta$$

$$= |c|^2 \left(\frac{a}{b} H(f(0)) + \frac{(\bar{a} - \bar{b})(1 - 2|b|^2 + a\bar{b}|b|^2)}{\bar{b}(1 - |b|^2)^2} H(f(1/\bar{b}))\right).$$

In short,

$$\frac{1}{\pi} \int_{\mathbb{C}} H v_f dm = AH(f(0)) + BH(f(1/\bar{b})), \tag{8.10}$$

where

$$A = |c|^2 \frac{a}{b}, \quad B = |c|^2 \frac{(\bar{a} - \bar{b})(1 - 2|b|^2 + a\bar{b}|b|^2)}{\bar{b}(1 - |b|^2)^2}. \tag{8.11}$$

For the harmonic moments (with respect to the weight v_f) this gives

$$M_0 = A + B,$$

$$M_k = Bf(1/\bar{b})^k, \quad k = 1, 2, \dots.$$

Here only M_0, M_1, M_2 are needed since the M_k lie in geometric progression from $k = 1$ on. From these three moments, A, B and $f(1/\bar{b})$ can be determined provided $M_1 \neq 0$, and after that a, b, c can be found, at least generically. Thus the moments M_0, M_1, M_2 suffice to locally determine f, and we can write

$$f(\zeta) = f(\zeta, M_0, M_1, M_2),$$

provided f is known *a priori* to be of the form (8.7) with $M_1 \neq 0$. However, as will be seen below, when we specialize to the case $M_1 = 0$ things change.

The rescue then is to introduce the branch point B_1 in (8.9) as an additional parameter. This will however give in total 7 real parameters for a family of functions which depends on only 5 parameters, so we should sacrifice two of them. Generically M_2 can be expected to be superfluous, resulting in the presentation

$$f(\zeta) = f(\zeta, M_0, M_1, B_1).$$

The quadrature Riemann surface picture enters when one starts from the second member in the computation leading to (8.10) and considers $h(\zeta) = H(f(\zeta))$ as an independent test function on the Riemann surface, thus allowing $h(\zeta_1) \neq h(\zeta_2)$ even when $f(\zeta_1) = f(\zeta_2)$. The quadrature identity becomes

$$\frac{1}{2\pi i} \int_{\mathbb{D}} h(\zeta) |f'(\zeta)|^2 d\bar{\zeta} d\zeta = Ah(0) + Bh(1/\bar{b}) \tag{8.12}$$

for h analytic and integrable (with respect to the weight $|f'|^2$) in the unit disk.

8.4.2   First Subcase

There are two cases which are of particular interest. These represent instances of $M_1 = 0$. The first case is when $a = 1/\bar{b}$. This does not change (8.12) very much, it is only that the two weights become equal:

$$\frac{1}{2\pi i} \int_{\mathbb{D}} h(\zeta) |f'(\zeta)|^2 d\bar{\zeta} d\zeta = Ah(0) + Ah(1/\bar{b}),$$

where $A = B = |c|^2/|b|^2$ and the normalization for c becomes $c > 0$. However, (8.10) changes more drastically because the two quadrature nodes now lie over the same point in the z-plane. Indeed $f(1/\bar{b}) = 0 = f(0)$, so (8.10) effectively becomes a one point identity:

$$\frac{1}{\pi} \int_{\mathbb{C}} Hv_f dm = 2AH(0)$$

and, for the moments,

$$M_0 = 2A, \quad M_1 = M_2 = \cdots = 0. \tag{8.13}$$

Clearly knowledge of these are not enough to determine f. This function originally had 5 real degrees of freedom. Two of them were used in condition $a = 1/\bar{b}$, but there still remain three, and $M_0 = 2A$ is only one real equation.

So something more would be needed, for example knowledge of the location of the branch point $B_1 = f(\omega_1)$. We have

$$\omega_1 = b\left(1 - \sqrt{1 - \frac{1}{|b|^2}}\right), \quad c = |b|\sqrt{\frac{M_0}{2}},$$

by which

$$B_1 = b|b|\sqrt{\frac{M_0}{2}}\left(1 - \sqrt{1 - \frac{1}{|b|^2}}\right)^2. \tag{8.14}$$

This equation can be solved for b in terms of B_1 and M_0. Indeed, by some elementary calculations one finds that

$$b = \frac{B_1}{2|B_1|}\left((\frac{2|B_1|^2}{M_0})^{1/4} + (\frac{2|B_1|^2}{M_0})^{-1/4}\right),$$

and after substitution of $a = 1/\bar{b}$, b and c one then has f explicitly on the form

$$f(\zeta) = f(\zeta, M_0, B_1).$$

By (8.13) the moment sequence is the same as that for the disk $\mathbb{D}(0, \sqrt{2A})$. One way to understand that is to observe that

$$f(\zeta) = -\frac{c}{\bar{b}} \cdot \zeta \cdot \frac{1 - \bar{b}\zeta}{\zeta - b}$$

is a function which maps $\mathbb{D}$ onto the disk $\mathbb{D}(0, \sqrt{A})$ covered twice. In other words, $\nu_f = 2\chi_{\mathbb{D}(0,\sqrt{A})}$. Note that the disk $\mathbb{D}(0, \sqrt{A}) = \mathbb{D}(0, \sqrt{M_0/2})$ depends only on M_0, not on B_1, so varying just B_1 keeps $f(\partial\mathbb{D})$ fixed as a set.

8.4.3 Second Subcase

The second interesting case is when $\omega_1 = 1/\bar{b}$. This means that the quadrature node $1/\bar{b}$ is at the same time a branch point. What also happens is that the quadrature node looses its weight: one gets $B = 0$ in (8.11). The quadrature node is still there, but it is only "virtual". (In principle it can be restored by allowing meromorphic test functions with a pole at the point, as the weight $|f'|^2$ certainly allows, but we shall not take such steps.) Thus we have again a one node quadrature identity, but this time in a more true sense, namely that it is such a quadrature identity on the

Riemann surface itself:

$$\frac{1}{2\pi i} \int_{\mathbb{D}} h(\zeta)|f'(\zeta)|^2 d\bar{\xi}\, d\zeta = Ah(0),\tag{8.15}$$

where

$$A = 2\frac{|c|^2}{|b|^2} - \frac{|c|^2}{|b|^4} = |c|^2|\omega_1|^2(2 - |\omega_1|^2).$$

Of course we also have

$$\frac{1}{\pi} \int_{\mathbb{C}} H v_f dm = AH(0),$$

and

$$M_0 = A, \quad M_1 = M_2 = \cdots = 0.\tag{8.16}$$

Here again the moments do not suffice to identify f. Indeed, we have now a one parameter family of functions f satisfying (8.16) with the same value of A. Explicitly this becomes, in terms of $\omega_1 = 1/\bar{b}$, which we keep as the free parameter,

$$f(\zeta) = C \cdot \frac{\zeta(2|\omega_1|^2 - |\omega_1|^4 - \bar{\omega}_1\zeta)}{1 - \bar{\omega}_1\zeta},$$

and where

$$C = \frac{\sqrt{M_0}}{|\omega_1|\sqrt{2 - |\omega_1|^2}}.$$

The branch point is

$$B_1 = f(\omega_1) = \frac{\omega_1|\omega_1|\sqrt{M_0}}{\sqrt{2 - |\omega_1|^2}}.$$

Since $|\omega_1| < 1$ we have $|B_1| < \sqrt{M_0}$. The above relationship can be inverted to give ω_1 in terms of B_1 and M_0:

$$\omega_1 = \frac{B_1}{|B_1|}\sqrt{-\frac{|B_1|^2}{2M_0} + \sqrt{\frac{|B_1|^4}{4M_0^2} + \frac{2|B_1|^2}{M_0}}}.$$

Thus one can explicitly write f on the form $f(\zeta) = f(\zeta, M_0, B_1)$ also in the present case.

It is interesting to also compute $f'(\zeta)$. One gets

$$f'(\zeta) = C \cdot \frac{(\zeta - \omega_1)(\zeta - \frac{2}{\omega_1} + \omega_1)}{(\zeta - \frac{1}{\bar{\omega}_1})^2},$$

which is, up to a constant factor, the contractive zero divisor in Bergman space corresponding to the zero $\omega_1 \in \mathbb{D}$, alternatively, the reproducing kernel for those square integrable analytic functions in $\mathbb{D}$ which vanish at ω_1. See [48, 49] for these concepts in general. Part of the meaning in the present case is simply that (8.15) holds. When ω_1 is real the above expression for f' is the same as (4.17).

8.5   Moment Evolutions in Terms of Poisson Brackets

Recall from (2.9) the general Poisson bracket

$$\{f_1, f_2\}_t = \zeta \frac{\partial f_1}{\partial \zeta} \frac{\partial f_2}{\partial t} - \zeta \frac{\partial f_2}{\partial \zeta} \frac{\partial f_1}{\partial t}. \tag{8.17}$$

This enters naturally when differentiating the formula (8.2) for the moments M_k with respect to t for a given evolution $f(\zeta, t)$. For a more general version we replace the time dependent function $f(\zeta, t)^k$ appearing in (8.2) by a function $h(\zeta, t)$ which is analytic in ζ and depends on t in the same way as $f(\zeta, t)^k$ does. This means that $h = h(\zeta, t)$ has to satisfy (4.23), that is

$$\frac{\partial h(\zeta, t)}{\partial t} \frac{\partial f(\zeta, t)}{\partial \zeta} = \frac{\partial f(\zeta, t)}{\partial t} \frac{\partial h(\zeta, t)}{\partial \zeta}, \tag{8.18}$$

and therefore locally can be regarded as a time independent function in the image domain of f.

A further generalization, which will be needed when differentiating with respect to higher order moments, is to allow t in (8.17) to be a complex variable, say $t = \tau + i\sigma$. Then $\frac{\partial}{\partial t}$ is interpreted as the complex Wirtinger derivative $\frac{\partial}{\partial t} = \frac{1}{2}(\frac{\partial}{\partial \tau} - i\frac{\partial}{\partial \sigma})$. We shall then also take the derivative $\frac{\partial}{\partial \bar{t}} = \frac{1}{2}(\frac{\partial}{\partial \tau} + i\frac{\partial}{\partial \sigma})$ into account and consider the corresponding Poisson bracket

$$\{f_1, f_2\}_{\bar{t}} = \zeta \frac{\partial f_1}{\partial \zeta} \frac{\partial f_2}{\partial \bar{t}} - \zeta \frac{\partial f_2}{\partial \zeta} \frac{\partial f_1}{\partial \bar{t}}. \tag{8.19}$$

The relevant version of (8.18) then is

$$\frac{\partial h(\zeta, t)}{\partial \bar{t}} \frac{\partial f(\zeta, t)}{\partial \zeta} = \frac{\partial f(\zeta, t)}{\partial \bar{t}} \frac{\partial h(\zeta, t)}{\partial \zeta}, \tag{8.20}$$

and the two Poisson brackets enjoy the symmetry

$$(\{f, f^*\}_t)^* = \{f, f^*\}_{\bar{t}}. \tag{8.21}$$

The variables t and $\bar{t}$ will be treated as independent variables. One may for example think of τ and σ in $t, \bar{t} = \tau \pm i\sigma$ as being themselves independent complex variables, which are eventually restricted to real values.

We now reformulate and extend Lemma 3.1 as

Lemma 8.1 *Assume that $h(\zeta, t)$ is analytic in ζ in a neighborhood of the closed unit disk and depends smoothly on a real or complex parameter t in such a way that (8.18), (8.20) hold (in the respective cases below). Then*

$$\frac{\partial}{\partial t} \int_{\mathbb{D}} h(\zeta, t) |f'(\zeta, t)|^2 d\bar{\zeta} d\zeta = \int_{\partial \mathbb{D}} h(\zeta, t) \{f, f^*\}_t \frac{d\zeta}{\zeta}, \tag{8.22}$$

$$\frac{\partial}{\partial \bar{t}} \int_{\mathbb{D}} h(\zeta, t) |f'(\zeta, t)|^2 d\bar{\zeta} d\zeta = \int_{\partial \mathbb{D}} h(\zeta, t) \{f, f^*\}_{\bar{t}} \frac{d\zeta}{\zeta}. \tag{8.23}$$

Proof The proofs are straight-forward: differentiating under the integral sign and using partial integration we have, for (8.22),

$$\frac{\partial}{\partial t} \int_{\mathbb{D}} h|f'|^2 d\bar{\zeta} d\zeta = \frac{\partial}{\partial t} \int_{\partial \mathbb{D}} h f^* f' d\zeta = \int_{\partial \mathbb{D}} \left(\dot{h} f^* f' + h \dot{f}^* f' + h f^* \dot{f}' \right) d\zeta$$

$$= \int_{\partial \mathbb{D}} \left(\dot{h} f^* f' + h \dot{f}^* f' - h' f^* \dot{f} - h(f^*)' \dot{f} \right) d\zeta$$

$$= \int_{\partial \mathbb{D}} \left((\dot{h} f' - \dot{f} h') f^* + h(\dot{f}^* f' - (f^*)' \dot{f}) \right) d\zeta = \int_{\partial \mathbb{D}} h \cdot \{f, f^*\}_t \frac{d\zeta}{\zeta},$$

where (8.18) was used in the last step. Similarly for (8.23). □

In the special case of $h(\zeta, t) = f(\zeta, t)^n, n \geq 0$, one gets

$$\frac{\partial}{\partial t} M_n = \frac{1}{2\pi i} \int_{\partial \mathbb{D}} f(\zeta, t)^n \{f, f^*\}_t \frac{d\zeta}{\zeta}. \tag{8.24}$$

Let, for a fixed number $k \geq 0$,

$$L_k(\zeta; f) = \frac{c_k}{\zeta^{k+1}} + \frac{c_{k-1}}{\zeta^k} + \cdots + \frac{c_1}{\zeta^2} + \frac{c_0}{\zeta} = \left(\frac{f'(\zeta)}{f(\zeta)^{k+1}} \right)_< \tag{8.25}$$

denote the singular part of of the Laurent series of $\frac{f'(\zeta)}{f(\zeta)^{k+1}}$ at $\zeta = 0$. The coefficients c_j depend on both f and k. Consider an evolution $t \mapsto f(\cdot, t)$ (with t possibly complex) for which

$$\{f, f^*\}_t = \zeta L_k(\zeta; f). \tag{8.26}$$

The parameter t here is linked to the number k. For such an evolution we have, by (8.22) and (8.24),

$$\frac{\partial}{\partial t} M_n = \frac{1}{2\pi i} \int_{\partial \mathbb{D}} f(\zeta)^n L_k(\zeta; f) d\zeta = \operatorname*{Res}_{\zeta=0} f(\zeta)^n \cdot \frac{f'(\zeta) d\zeta}{f(\zeta)^{k+1}} = \begin{cases} 1 & \text{if } n = k, \\ 0 & \text{if } n \neq k. \end{cases}$$

For the same evolution, the symmetry (8.21) shows that

$$\{f, f^*\}_{\bar{t}} = \frac{1}{\zeta} L_k^*(\zeta; f) = \bar{c}_0 + \bar{c}_1 \zeta + \cdots + \bar{c}_k \zeta^k.$$

Therefore

$$\frac{\partial}{\partial \bar{t}} M_n = \frac{1}{2\pi i} \int_{\partial \mathbb{D}} f(\zeta, t)^n \{f, f^*\}_{\bar{t}} \frac{d\zeta}{\zeta} = f(0, t)^n \bar{c}_0,$$

which is zero unless $n = 0$. Thus

$$\frac{\partial}{\partial \bar{t}} \overline{M}_n = \overline{\frac{\partial}{\partial \bar{t}} M_n} = 0$$

when $n > 0$. Since

$$c_0 = \operatorname{Res} \frac{f'(\zeta) d\zeta}{f(\zeta)^{k+1}} = \begin{cases} 0 & \text{when } k > 0, \\ 1 & \text{when } k = 0, \end{cases}$$

we conclude (again, since $\overline{M}_0 = M_0$) that

$$\frac{\partial \overline{M}_0}{\partial t} = \begin{cases} 0 & \text{when } k > 0, \\ 1 & \text{when } k = 0. \end{cases}$$

When (8.26) holds with $k > 0$, then t is necessarily a complex variable. Therefore $\frac{\partial}{\partial \bar{t}} \overline{M}_n = 0$ is not in conflict with $\frac{\partial}{\partial t} M_n = 1$. Indeed, M_n and $\overline{M}_n$ are in the present context independent variables, as indicated in Example 8.1. In conclusion, we have for any $k \geq 0$ constructed an evolution with a complex evolution parameter t such that, effectively, $t = t_k = M_k$ (+constant).

The arguments can easily be run in the other direction, hence we can summarize as follows.

Theorem 8.1 *For any fixed $n \geq 1$, and with L_n defined by (8.25), the differential equation (8.26), i.e.*

$$\{f, f^*\}_t = \zeta L_n(\zeta; f), \tag{8.27}$$

characterizes those evolutions $t \mapsto f(\cdot, t)$ for which M_n increases linearly in time (with speed one) and all remaining moments are conserved. The corresponding equation for which only $\overline{M}_n$ changes (with speed one) is

$$\{f, f^*\}_t = \frac{1}{\zeta} L_n^*(\zeta; f).$$ (8.28)

When $n = 0$, the equation for change of only M_0 (with speed one) is

$$\{f, f^*\}_t = 1.$$

Equation (8.27) can be solved for $\frac{\partial f}{\partial t}$ to give an equation of Löwner-Kufarev type, exactly as was done in Chap. 2 in the steps leading from (2.1) to (2.10). This gives an expression for $\frac{\partial f}{\partial M_n}$ in terms of a Poisson integral. That equation then guarantees that also the branch points remain fixed under the evolution. Compare Theorem 2.1. In the next chapter we shall derive formulas for $\frac{\partial f}{\partial M_n}$ within a Hamiltonian framework.

Chapter 9
Hamiltonian Descriptions of General Laplacian Evolutions

Abstract We formulate the dependence of the conformal map on harmonic moments and other parameters within a Hamiltonian framwork and by systematically using differential forms and other concept from differential geometry. This gives alternative ways of expressing Poisson brackets and the string equation.

9.1 Lie Derivatives and Interior Multiplication

Equation (8.22) and the Poisson bracket therein are implicitly related to concepts like the Lie derivative and interior multiplication, used in general differential geometry for handling differentiation of integrals with moving domains of integration. Here we shall make this relationship more explicit. As for notations within differential geometry we mostly follow [22].

In order to apply the mentioned tools in situations of Laplacian evolutions we need to impose stronger assumptions on regularity of functions and boundaries than one usually do in differential geometry. In general boundaries of domains are assumed to be real analytic and functions defined up to these boundaries to have real analytic continuations across them. Such assumptions are necessary because of fundamental properties of Laplacian growth: an evolution governed by for example (9.7), (9.8) (see below) does not exist in both time directions unless $\partial\Omega(t)$ is analytic. Then, in this particular case, the harmonic function p there will automatically have a harmonic extension across $\partial\Omega(t)$.

Now, let $\mathbf{V}(z, t)$ quite generally be a, possibly time dependent (with real valued time t), vector field in the complex plane, and let $\Omega(t)$ be a family of domains which moves in the flow of this field. Then, if $\Phi(z)$ is a function of the space variables the time derivative of the integral of Φ can be obtained as

$$\frac{d}{dt}\int_{\Omega(t)}\Phi(z)dx \wedge dy = \int_{\Omega(t)} L_{\mathbf{V}}(\Phi\, dx \wedge dy),$$

where L_V denotes the **Lie derivative**, see in general [22, 122], and where area measure $dxdy = dx \wedge dy$ is treated as a differential form. When acting on a differential form the Lie derivative is given by the homotopy formula

$$L_V = d \circ i(\mathbf{V}) + i(\mathbf{V}) \circ d,$$

where $i(\mathbf{V})$ denotes **interior multiplication** ("contraction") by $\mathbf{V}$. Since $d(\Phi \, dx \wedge dy) = 0$ we get in our case, using Stokes' formula,

$$\frac{d}{dt} \int_{\Omega(t)} \Phi \, dx \wedge dy = \int_{\Omega(t)} d(i(\mathbf{V})(\Phi \, dx \wedge dy))$$

$$= \int_{\partial\Omega(t)} i(\mathbf{V})(\Phi dx \wedge dy) = \int_{\partial\Omega(t)} \Phi(i(\mathbf{V})(dx \wedge dy)).$$

Writing $\mathbf{V}$ in terms of components as

$$\mathbf{V} = V_x \frac{\partial}{\partial x} + V_y \frac{\partial}{\partial y} = V_z \frac{\partial}{\partial z} + V_{\bar{z}} \frac{\partial}{\partial \bar{z}} \tag{9.1}$$

we have

$$i(\mathbf{V})(dx \wedge dy) = V_x dy - V_y dx = i(\mathbf{V})(\frac{i}{2} dz \wedge d\bar{z}) = \frac{i}{2}(V_z d\bar{z} - V_{\bar{z}} dz).$$

The basis vectors $\frac{\partial}{\partial x}, \frac{\partial}{\partial y}$ in the real tangent space of $\mathbb{C}$ can in our setting be identified with the complex numbers 1 and i, respectively. If $\mathbf{V}$ is a vector in this real tangent space, then $V_x = \mathrm{Re}\,\mathbf{V}$, $V_y = \mathrm{Im}\,\mathbf{V}$ are real, while V_z, $V_{\bar{z}}$ are complex conjugates of each other. At any point on $\partial\Omega$ one may also introduce an orthonormal basis $\frac{\partial}{\partial n}$, $\frac{\partial}{\partial s}$ representing the (outward) normal and tangential directions (s thought of as arc length), and correspondingly write

$$\mathbf{V} = V_n \frac{\partial}{\partial n} + V_s \frac{\partial}{\partial s}.$$

In terms of this we have

$$i(\mathbf{V})(dx \wedge dy) = i(\mathbf{V})(dn \wedge ds) = V_n \, ds - V_s \, dn.$$

Since $dn = 0$ when integrating along the boundary we get the natural formula

$$\frac{d}{dt} \int_{\Omega(t)} \Phi \, dxdy = \int_{\partial\Omega(t)} \Phi \, V_n ds,$$

which has already been used several times.

Next we adapt the above machinery to our situation, with time dependent conformal maps, which we assume to be univalent for the time being:

$$f(\cdot, t) : \mathbb{D} \to \Omega(t).$$

These conformal maps may be viewed simply as changes of coordinates, and we shall pull back the above formulas to $\mathbb{D}$. The vector field $\mathbf{V}$ itself is, when viewed as a complex number via the identifications $\frac{\partial}{\partial x} \leftrightarrow 1, \frac{\partial}{\partial y} \leftrightarrow i$,

$$\mathbf{V}(z, t) = \dot{f}(\zeta, t), \quad \text{where } z = f(\zeta, t). \tag{9.2}$$

More precisely this means that $\mathbf{V}$ is the velocity of the point $z = f(\zeta, t)$ with $\zeta \in \partial\mathbb{D}$ kept fixed. In particular, $\mathbf{V}$ will in general not be perpendicular to the boundary. As a complex number the derivative $\dot{f}$ in (9.2) is also the same as V_z in (9.1), and $\dot{\bar{f}}$ is $V_{\bar{z}}$. Hence, as is confirmed by spelling out the Wirtinger derivatives:

$$\mathbf{V} = \dot{f} \frac{\partial}{\partial z} + \dot{\bar{f}} \frac{\partial}{\partial \bar{z}}.$$

On pulling back to $\mathbb{D}$ we have

$$dx \wedge dy = \frac{i}{2} dz \wedge d\bar{z} = \frac{i}{2} df \wedge d\bar{f},$$

$$i(\mathbf{V})(dz \wedge d\bar{z}) = i(\dot{f} \frac{\partial}{\partial z} + \dot{\bar{f}} \frac{\partial}{\partial \bar{z}})(dz \wedge d\bar{z})$$

$$= \dot{f} \, d\bar{z} - \dot{\bar{f}} \, dz = \dot{f} \, d\bar{f} - \dot{\bar{f}} \, df.$$

Next we want to incorporate time t more systematically as one of the coordinates, on more or less equal footing as (x, y), i.e. we go from "phase space" to "extended phase space" in a Hamiltonian mechanics terminology. In this extended space the original complex z-plane is represented by the surface

$$\mathscr{S} = \{(x, y, t) : x + iy \in \partial\Omega(t)\}, \tag{9.3}$$

or by the cylinder $\partial\mathbb{D} \times \mathbb{R}$ when pulled back by $z = f(\zeta, t)$. These are two dimensional manifolds, and near any point where $\mathbf{V} \neq 0$ the variables x and y (or simply $z = x + iy$) can be used as local coordinates on $\mathscr{S}$. And on $\partial\mathbb{D} \times \mathbb{R}$ it is natural to use (θ, t), where $f(e^{i\theta}, t) = x + iy$. Mostly we shall write (ζ, t) however, where it is understood that $\zeta \in \partial\mathbb{D}$ when a point on the surface $\mathscr{S}$ is represented.

In terms of the (ζ, t)-coordinates, the complete differentials of f and $\bar{f}$ are

$$df = f'd\zeta + \dot{f}dt, \quad d\bar{f} = \overline{f'}d\bar{\zeta} + \dot{\bar{f}}dt,$$

On the cylinder $\partial \mathbb{D} \times \mathbb{R}$, the variables ζ and $\bar{\zeta}$ are coupled by $\zeta \bar{\zeta} = 1$, so that in particular $d\zeta \wedge d\bar{\zeta} = 0$. It follows that

$$df \wedge d\bar{f} = (\zeta f' \, \dot{\bar{f}} + \bar{\zeta} \, \overline{f'} \, \dot{f}) \frac{d\zeta}{\zeta} \wedge dt. \tag{9.4}$$

In terms of f^*, the holomorphic reflection of f, we have $df^* = (f^*)' d\zeta + \dot{f}^* dt$, and the relation (9.4) becomes

$$df \wedge df^* = \{f, f^*\}_t \frac{d\zeta}{\zeta} \wedge dt. \tag{9.5}$$

Alternatively, in terms of the action of a 2-form on a pair of vectors:

$$\{f, f^*\}_t = (df \wedge df^*)(\zeta \frac{\partial}{\partial \zeta}, \frac{\partial}{\partial t}). \tag{9.6}$$

It is advantageous to work with f^*, because it is holomorphic. When a relation involving f and f^* holds on $\partial \mathbb{D} \times \mathbb{R}$, then it automatically extends holomorphically to a full neighborhood of $\partial \mathbb{D} \times \mathbb{R}$ in $\mathbb{D} \times \mathbb{R}$ when expressed in terms of f and f^*. In this way ζ and t become free and independent variables, despite they are in total three (real) variables for a relation holding on a two dimensional manifold. The derivation $\frac{\partial}{\partial t}$ can be regarded as a vector field on $\mathscr{S}$ via its action on a test function φ on $\mathscr{S}$ by

$$\frac{\partial \varphi}{\partial t} = \frac{\partial \varphi}{\partial x} \frac{\partial x}{\partial t} + \frac{\partial \varphi}{\partial y} \frac{\partial y}{\partial t} = V_x \frac{\partial \varphi}{\partial x} + V_y \frac{\partial \varphi}{\partial y} = (\mathbf{V} \cdot \nabla)\varphi.$$

In this sense $\frac{\partial}{\partial t}$ represents $\mathbf{V}$ as a directional derivative. Also on $\partial \mathbb{D} \times \mathbb{R}$ does $\frac{\partial}{\partial t}$ represent $\mathbf{V}$, as is clear from (9.2).

9.2   Laplacian Evolutions

The next step is to look at all this in the context of general Laplacian evolutions. This means that we assume that the normal component V_n of the velocity on the boundary arises as the normal derivative of a function $p = p(z, t)$ which vanishes on $\partial \Omega(t)$ and is harmonic in $\Omega(t)$, except for some given singularities. This function p may for example be some derivative, with respect to $a \in \Omega$, of the Green function $G(z, a) = -\log|z - a| +$ harmonic with pole at a, and with the point a eventually chosen (in our applications) to be $a = 0$.

Thus the dynamics is determined, given the singularities for p, by

$$p(\cdot, t) = 0 \quad \text{on } \partial\Omega(t), \tag{9.7}$$

$$-\frac{\partial p(\cdot, t)}{\partial n} = V_n \quad \text{on } \partial\Omega(t). \tag{9.8}$$

We assume that the boundary really moves, more precisely that $V_n \neq 0$ at any point of consideration. There are three natural sets of local coordinates on the surface $\mathscr{S}$ (see (9.3)). One is the Cartesian coordinates (x, y) (or $z = x + iy$) of the projection of $\mathscr{S}$ onto the original complex plane. A second consists of an arc length parameter s along $\partial\Omega(t)$, for a fixed t, together with a normal coordinate n along the outward normal. The latter is far from being uniquely determined, but we need to consider it only on the tangent space at a fixed point, and then it is to be chosen so that the frame $(\frac{\partial}{\partial n}, \frac{\partial}{\partial s})$ is just a rotation of the Cartesian frame $(\frac{\partial}{\partial x}, \frac{\partial}{\partial y})$. Then

$$dn \wedge ds = dx \wedge dy.$$

The third set of coordinates are derived from (9.7), (9.8). The latter equation says that $*dp = -V_n \, ds$ along $\partial\Omega$. Here the star is the Hodge star, and this is compatible with the star denoting harmonic conjugate, which we write on the left hand side of the function symbol to avoid confusion with the star which denotes holomorphic reflection. So $*dp = d(*p)$, where $*p$ is the harmonic conjugate of p. This is not uniquely determined, but locally one can always fix a choice by some normalization. In the sequel we assume this has been done. The Cauchy-Riemann equations connecting p and $*p$ give, when expressed in the (n, s)-coordinates, that

$$\frac{\partial(*p)}{\partial s} = \frac{\partial p}{\partial n} = -V_n$$

at any point of $\partial\Omega$ under consideration.

This shows that $*p$ can be viewed as a coordinate which increases with speed $-V_n$ along any fixed $\partial\Omega(t)$. A complementary orthogonal coordinate is then t itself, as it appears in (9.3). This increases in the normal direction with speed

$$\frac{dt}{dn} = \frac{1}{V_n},$$

since $\frac{dn}{dt} = V_n$, informally speaking. Thus we have the "string equation"

$$*dp \wedge dt = dn \wedge ds = dx \wedge dy, \tag{9.9}$$

expressing unit Jacobi determinant when changing between the mentioned coordinates on $\mathscr{S}$.

Using complex variables we can write

$$*dp \wedge dt = \frac{i}{2} dz \wedge d\bar{z},$$

and since $dp = 0$ along $\partial\Omega(t)$ it is possible to add any multiple of dp in the left member, leading specifically to

$$2dt \wedge (dp + i * dp) = dz \wedge d\bar{z}.$$

The above equations hold on $\mathscr{S}$, and when pulling back everything to (ζ, t)-space we get analogous equations holding on the cylinder $\partial\mathbb{D} \times \mathbb{R}$, for example

$$2dt \wedge d\mathscr{H}_t = df \wedge d\bar{f}.$$

Here we have set

$$\mathscr{H}_t = \left(p + i(^*p)\right) \circ f, \tag{9.10}$$

and consider this as a **Hamiltonian function** associated to the particular evolution parametrized by t.

Now with both $\mathscr{H}_t$ and f holomorphic it is natural to write all equations as holomorphic equations, then holding identically with ζ and t as free variables in a neighborhood of $\partial\mathbb{D} \times \mathbb{R}$:

$$2dt \wedge d\mathscr{H}_t = df \wedge df^*. \tag{9.11}$$

As a consequence, in view of (9.5), we have

$$\{f, f^*\}_t = -2\zeta \frac{\partial\mathscr{H}_t}{\partial\zeta}. \tag{9.12}$$

The above considerations were of a general nature but, quite remarkably, one can still say a lot about the structure of the Hamiltonians. Since p vanishes on $\partial\Omega$, and is harmonic with certain singularities in Ω, it follows that p extends across $\partial\Omega$ to an odd function on the **Schottky double** $\hat{\Omega}$ of Ω, that is the compact Riemann surface obtained by completing Ω with a backside, identical with Ω as a set but having the opposite conformal structure, the two surfaces being welded along $\partial\Omega$. Also the harmonic conjugate $^*p$ extends, as an even function. It may happen that this becomes multi-valued (actually already in Ω), as is the case when $p = G$ is the Green function itself, so it is better to say that the differentials dp and $*dp$ extend to the Schottky double. The conclusion is that $dp + i*dp$ is a meromorphic differential (Abelian differential) on the Schottky double.

Example 9.1 If p is a derivative (acting on the pole) of order $n \geq 1$ of the Green function at $a = 0$, say

$$p(z) = \frac{1}{n!} \operatorname{Re} \frac{\partial^n}{\partial a^n} \big|_{a=0} G(z, a) = \operatorname{Re} \left(\frac{1}{2n z^n} + \text{holomorphic} \right),$$

then

$$(dp + i * dp)(z) = -\frac{dz}{2z^{n+1}} + \text{holomorphic}, \tag{9.13}$$

in Ω, and similarly on the reverse side. In other words, $dp + i * dp$ is an Abelian differential on the Schottky double with an $(n + 1)$:th order pole at $z = 0$ and a corresponding pole at the mirror point of $z = 0$.

When pulled back to $\mathbb{D}$ the conclusion is that, in a neighborhood of $\zeta = 0$, the differential $d\mathcal{H}_t$, with respect to ζ (only), looks like

$$d\mathcal{H}_t = -\frac{f'(\zeta, t)d\zeta}{2f(\zeta, t)^{n+1}} + \text{holomorphic}.$$

Similarly in $\mathbb{D}^e$, with a reflected singularity at infinity. In brief, $\frac{\partial \mathcal{H}_t}{\partial \zeta}$ is a rational function with poles of order $n + 1$ at $\zeta = 0$ and $\zeta = \infty$. This applies also when $n = 0$, then with $p = G$.

Inserting this information into (9.12) gives, in the notation of (8.25),

$$\{f, f^*\}_t = \zeta L_n(\zeta; f) + \frac{1}{\zeta} L_n^*(\zeta; f). \tag{9.14}$$

Here the right member is symmetric under holomorphic involution in $\partial \mathbb{D}$, which is necessary for the time variable t to be real. Indeed, we have

$$\zeta L_n(\zeta; f) + \frac{1}{\zeta} L_n^*(\zeta; f) = \frac{c_n}{\zeta^n} + \cdots + \frac{c_1}{\zeta} + c_0 + \bar{c}_0 + \bar{c}_1 \zeta + \cdots + \bar{c}_n \zeta^n.$$

Here $c_0 = 0$ when $n \geq 1$, $c_0 = 1$ when $n = 0$.

9.3 Schwarz Potentials and Generating Functions

Recall the definition (3.19) of the Schwarz function $S(z)$ of $\partial\Omega$ in terms of the potential u, the latter defined as the solution of the Cauchy problem (3.18). We shall presently take u to be the two-sided solution of (3.18), so that $\Delta u = 1$ in a full neighborhood of $\partial\Omega$, with $u = |\nabla u| = 0$ on $\partial\Omega$. Starting from this u one can define a harmonic potential w by

$$w(z) = \frac{1}{2}|z|^2 - 2u(z).$$

This satisfies

$$S(z) = 2\frac{\partial w}{\partial z}.$$

Let $^*w$ denote a (locally defined) harmonic conjugate function of w, so that $W = w + i^*w$ is analytic. Then

$$S(z) = \frac{\partial W(z)}{\partial z}. \tag{9.15}$$

Globally $W(z)$ will be multi-valued. For example for $\Omega = \mathbb{D}$ we have $S(z) = \frac{1}{z}$ and hence $W(z) = \log z$, plus an undetermined imaginary (and local) constant.

Under a Laplacian evolution $\Omega(t)$ driven by a distribution μ as in Sect. 3.4, see in particular (3.22), the potentials u, w and W depend on t in such a way that the time derivative of u equals that harmonic potential p which generates the evolution, see (3.21), (3.25). It follows that $\frac{\partial w}{\partial t} = -2p$ and, after analytic completion and adjustment of imaginary parts,

$$\frac{\partial W(z, t)}{\partial t} = -2(p + i^*p). \tag{9.16}$$

This **Schwarz potential** $W(z, t)$ can be pulled back to unit disk and then becomes a function of ζ and t, which we denote

$$V(\zeta, t) = W(f(\zeta, t), t). \tag{9.17}$$

In the non-univalent case this is the potential one has to work with. In view of (3.20), (9.15), (9.16) it satisfies

$$\frac{\partial V}{\partial \zeta} = f^*\frac{\partial f}{\partial \zeta}, \tag{9.18}$$

$$\frac{\partial V}{\partial t} = f^*\frac{\partial f}{\partial t} - 2\mathcal{H}_t, \tag{9.19}$$

where $\mathcal{H}_t$ is defined by (9.10). We can write (9.18), (9.19) as a single relation for the differentials:

$$dV = f^*df - 2\mathcal{H}_t\,dt.$$

In a Hamiltonian perspective this says that V is the generating function for a canonical transformation (see [5]) of variables in phase space, for example the transformation from $(\mathcal{H}_t, t)$ to (f, f^*). Compare discussions in [124]. The string equation on the form (9.11) expresses exactly the fact that dV is a closed one-form.

9.4 Multitime Hamiltonians

The previous section was partly of preparatory nature, for showing in detail how time can be properly integrated with the space variables in an extended phase space. Following ideas from (for example) [17, 62, 63, 80, 112, 124] we shall in this section let the single time variable t blow up to an infinite number of times $t_0, t_1, t_2, \ldots$, which will be the harmonic moments, augmented in the non-univalent case with other variables, like branch points. Here $t_0 = t$ is ordinary (real) time, while the $t_1, t_2, \ldots$ will be complex, with their complex conjugates $\bar{t}_1, \bar{t}_2, \ldots$ considered as independent variables. Notationally we shall however stick to $M_0, M_1, \bar{M}_1, \ldots$ (etc.).

So far we have mainly thought about single-sheeted domains Ω, or conformal maps f which are univalent. In the general case, with $f \in \mathcal{O}_{\mathrm{norm}}(\mathbb{D})$ assumed only to satisfy $f' \neq 0$ on $\partial \mathbb{D}$, equations such as (9.11) can be taken over more or less literally. Since the harmonic moments will usually not be enough to characterize Ω or f in this situation, additional variables are needed, and a natural choice for these are the branch points $B_1, B_2, \ldots$. Thus we write

$$f(\zeta) = f(\zeta, M_0, M_1, \ldots, B_1, B_2, \ldots),$$

or $f(\zeta, M_0, M_1, \bar{M}_1, \ldots, B_1, \bar{B}_1, \ldots)$, if we insist on exhibiting analytic dependence.

The surface $\mathscr{S}$ in (9.3) will now correspond to an infinite dimensional manifold, parametrized, as a replacement for $\partial \mathbb{D} \times \mathbb{R}$, by

$$\partial \mathbb{D} \times \mathbb{R} \times \mathbb{C} \times \mathbb{C} \times \ldots,$$

where the first two factors are for the variables ζ and M_0 and the remaining for the M_k ($k \geq 1$) and B_j. The differential of f with respect to all variables is, taking into account that those after M_0 are complex variables and that the dependence on these is not analytic,

$$df = \frac{\partial f}{\partial \zeta} d\zeta + \frac{\partial f}{\partial M_0} dM_0$$

$$+ \sum_{k \geq 1} \left(\frac{\partial f}{\partial M_k} dM_k + \frac{\partial f}{\partial \bar{M}_k} d\bar{M}_k \right)$$

$$+ \sum_{j \geq 1} \left(\frac{\partial f}{\partial B_j} dB_j + \frac{\partial f}{\partial \bar{B}_j} d\bar{B}_j \right).$$

Such differentials with infinitely many terms are to be interpreted formally, and no convergence aspects are involved. At any particular application one changes only finitely many variables, and the differentials of the remaining variables are then zero.

The Schwarz potential V in (9.17) becomes, like f, a function of ζ and all parameters, in short (with M and B shorthand for all the moments and branch points, respectively)

$$V = V(\zeta, M, B),$$

and the defining relations, corresponding to (9.18), (9.19), become

$$\frac{\partial V}{\partial \zeta} = f^* \frac{\partial f}{\partial \zeta}, \tag{9.20}$$

$$\frac{\partial V}{\partial M_0} = f^* \frac{\partial f}{\partial M_0} - \mathcal{H}_0, \tag{9.21}$$

$$\frac{\partial V}{\partial M_k} = f^* \frac{\partial f}{\partial M_k} - \mathcal{H}_k, \qquad \frac{\partial V}{\partial \bar{M}_k} = f^* \frac{\partial f}{\partial \bar{M}_k} - \tilde{\mathcal{H}}_k, \tag{9.22}$$

$$\frac{\partial V}{\partial B_j} = f^* \frac{\partial f}{\partial B_j} - \mathcal{K}_j, \qquad \frac{\partial V}{\partial \bar{B}_j} = f^* \frac{\partial f}{\partial \bar{B}_j} - \tilde{\mathcal{K}}_j \tag{9.23}$$

for appropriate Hamiltonians $\mathcal{H}_0, \mathcal{H}_k, \tilde{\mathcal{H}}_k, \mathcal{K}_j, \tilde{\mathcal{K}}_j$ $(k, j \geq 1)$, defined by the above equations. Equation (9.21) is in accordance with (9.19) since $dM_0 = 2dt$.

The differential of V is expanded similarly as that of f, and in view of the above relations this results in

$$dV = f^* df - \mathcal{H}_0 \, dM_0 - \sum_{k \geq 1} \left(\mathcal{H}_k \, dM_k + \tilde{\mathcal{H}}_k \, d\bar{M}_k \right) - \sum_{j \geq 1} \left(\mathcal{K}_j \, dB_j + \tilde{\mathcal{K}}_j \, d\bar{B}_j \right).$$

Since dV is closed we obtain the identity

$$df \wedge df^* = dM_0 \wedge d\mathcal{H}_0$$
$$+ \sum_{k \geq 1} \left(dM_k \wedge d\mathcal{H}_k + d\bar{M}_k \wedge d\tilde{\mathcal{H}}_k \right) + \sum_{j \geq 1} \left(dB_j \wedge d\mathcal{K}_j + d\bar{B}_j \wedge d\tilde{\mathcal{K}}_j \right).$$

$$\tag{9.24}$$

This expression gives handy formulas for the differentials of the Hamiltonians via interior multiplication (or interior derivation):

$$d\mathcal{H}_0 = i(\frac{\partial}{\partial M_0})(df \wedge df^*),$$

$$d\mathcal{H}_k = i(\frac{\partial}{\partial M_k})(df \wedge df^*), \qquad d\tilde{\mathcal{H}}_k = i(\frac{\partial}{\partial \bar{M}_k})(df \wedge df^*),$$

$$d\mathcal{H}_j = i(\frac{\partial}{\partial B_j})(df \wedge df^*), \qquad d\tilde{\mathcal{H}}_0 = i(\frac{\partial}{\partial \bar{B}_j})(df \wedge df^*). \qquad (9.25)$$

It is easy to see (compare (8.21)) that

$$(df \wedge df^*)^* = -df \wedge df^*,$$

hence we have the symmetries

$$d\tilde{\mathcal{H}}_k = -d\mathcal{H}_k^*, \quad d\tilde{\mathcal{H}}_j = -d\mathcal{H}_j^*.$$

Now we want to be find out as much as possible about the structure the Hamiltonians. For $\mathcal{H}_0$ one can use directly (9.10), applying this with t being the time variable in the original Polubarinova-Galin equation (2.1), or better the Löwner-Kufarev equation (2.10), with $q(t) = 1$. This was driven by the Green function, that is $p = G$, and all parameters M_k, B_j $(k, j \geq 1)$ are preserved under the evolution, which makes $\frac{\partial}{\partial t}$ have the meaning of $\frac{\partial}{\partial M_0}$, up to a factor. Thus Hamiltonian $\mathcal{H}_t$ in (9.10) equals, as a function of ζ, the analytic completion of this Green function pulled back to $\mathbb{D}$:

$$\mathcal{H}_t(\zeta) = ((G + i^*G) \circ f)(\zeta) = -\log \zeta.$$

On the other hand, the relation between t and M_0 is $2dt = dM_0$, see (2.5), where now $Q(t) = t$, hence on identifying the first term in (9.24) with (9.11) we conclude that $\mathcal{H}_0$ is the same as this $\mathcal{H}_t$:

$$\mathcal{H}_0 = -\log \zeta.$$

Here the imaginary part is multivalued, but locally one can always fix a branch of it. On the other hand one must allow this branch to depend on all parameters, so eventually one has to write

$$\mathcal{H}_0(\zeta) = -\log \zeta + b_0(M, B), \qquad (9.26)$$

where $b_0(M, B)$ is an imaginary constant (with respect to ζ) depending of the moments and branch points.

Spelling out (9.25) gives, after simplifications,

$$
d\mathcal{H}_0 = i\left(\frac{\partial}{\partial M_0}\right)(df \wedge df^*) = -\{f, f^*\}\frac{d\zeta}{\zeta} + \left(\frac{\partial f}{\partial M_0}\frac{\partial f^*}{\partial M_0} - \frac{\partial f^*}{\partial M_0}\frac{\partial f}{\partial M_0}\right)dM_0
$$

$$
+ \sum_{j\geq 1}\left(\frac{\partial f}{\partial M_0}\frac{\partial f^*}{\partial M_j} - \frac{\partial f^*}{\partial M_0}\frac{\partial f}{\partial M_j}\right)dM_j + \sum_{j\geq 1}\left(\frac{\partial f}{\partial M_0}\frac{\partial f^*}{\partial \bar{M}_j} - \frac{\partial f^*}{\partial M_0}\frac{\partial f}{\partial \bar{M}_j}\right)d\bar{M}_j
$$

$$
+ \sum_{j\geq 1}\left(\frac{\partial f}{\partial M_0}\frac{\partial f^*}{\partial B_j} - \frac{\partial f^*}{\partial M_0}\frac{\partial f}{\partial B_j}\right)dB_j + \sum_{j\geq 1}\left(\frac{\partial f}{\partial M_0}\frac{\partial f^*}{\partial \bar{B}_j} - \frac{\partial f^*}{\partial M_0}\frac{\partial f}{\partial \bar{B}_j}\right)d\bar{B}_j.
$$

Since the coefficient for dM_0 above vanishes, we conclude that the constant $b_0(M, B)$ in (9.26) does not depend on M_0. Hence

$$
\frac{\partial \mathcal{H}_0}{\partial M_0} = 0. \tag{9.27}
$$

We also see that

$$
\zeta\frac{\partial \mathcal{H}_0}{\partial \zeta} = -\{f, f^*\}. \tag{9.28}
$$

Since on the other hand, $\zeta\frac{\partial \mathcal{H}_0}{\partial \zeta} = -1$ by (9.26), we arrive again at the string equation

$$
\{f, f^*\} = 1. \tag{9.29}
$$

Similar computations apply to the Hamiltonians $\mathcal{H}_k$, $\tilde{\mathcal{H}}_k$, $\mathcal{K}_j$, $\tilde{\mathcal{K}}_j$. Considering for example $d\mathcal{H}_k$ with $k > 0$, obtained via interior multiplication of $df \wedge df^*$ by $\frac{\partial}{\partial M_k}$ as above, we have

$$
d\mathcal{H}_k = -\left(\frac{\partial f}{\partial \zeta}\frac{\partial f^*}{\partial M_k} - \frac{\partial f^*}{\partial \zeta}\frac{\partial f}{\partial M_k}\right)d\zeta - \left(\frac{\partial f}{\partial M_0}\frac{\partial f^*}{\partial M_k} - \frac{\partial f^*}{\partial M_0}\frac{\partial f}{\partial M_k}\right)dM_0
$$

$$
- \sum_{j\geq 1}\left(\frac{\partial f}{\partial M_j}\frac{\partial f^*}{\partial M_k} - \frac{\partial f^*}{\partial M_j}\frac{\partial f}{\partial M_k}\right)dM_j - \sum_{j\geq 1}\left(\frac{\partial f}{\partial \overline{M}_j}\frac{\partial f^*}{\partial M_k} - \frac{\partial f^*}{\partial \overline{M}_j}\frac{\partial f}{\partial M_k}\right)d\bar{M}_j
$$

$$
- \sum_{j\geq 1}\left(\frac{\partial f}{\partial B_j}\frac{\partial f^*}{\partial M_k} - \frac{\partial f^*}{\partial B_j}\frac{\partial f}{\partial M_k}\right)dB_j - \sum_{j\geq 1}\left(\frac{\partial f}{\partial \overline{B}_j}\frac{\partial f^*}{\partial M_k} - \frac{\partial f^*}{\partial \overline{B}_j}\frac{\partial f}{\partial M_k}\right)d\bar{B}_j.
$$

This shows that

$$-\frac{\partial \mathcal{H}_k}{\partial \zeta} = \frac{\partial f}{\partial \zeta}\frac{\partial f^*}{\partial M_k} - \frac{\partial f}{\partial M_k}\frac{\partial f^*}{\partial \zeta}, \tag{9.30}$$

$$-\frac{\partial \mathcal{H}_k}{\partial M_0} = \frac{\partial f}{\partial M_0}\frac{\partial f^*}{\partial M_k} - \frac{\partial f}{\partial M_k}\frac{\partial f^*}{\partial M_0}, \tag{9.31}$$

$$-\frac{\partial \mathcal{H}_k}{\partial M_j} = \frac{\partial f}{\partial M_j}\frac{\partial f^*}{\partial M_k} - \frac{\partial f}{\partial M_k}\frac{\partial f^*}{\partial M_j}, \quad j \geq 1. \tag{9.32}$$

$$-\frac{\partial \mathcal{H}_k}{\partial \bar{M}_j} = \frac{\partial f}{\partial \bar{M}_j}\frac{\partial f^*}{\partial M_k} - \frac{\partial f}{\partial M_k}\frac{\partial f^*}{\partial \bar{M}_j}, \quad j \geq 1. \tag{9.33}$$

$$-\frac{\partial \mathcal{H}_k}{\partial B_j} = \frac{\partial f}{\partial B_j}\frac{\partial f^*}{\partial M_k} - \frac{\partial f}{\partial M_k}\frac{\partial f^*}{\partial B_j}, \quad j \geq 1. \tag{9.34}$$

$$-\frac{\partial \mathcal{H}_k}{\partial \bar{B}_j} = \frac{\partial f}{\partial \bar{B}_j}\frac{\partial f^*}{\partial M_k} - \frac{\partial f}{\partial M_k}\frac{\partial f^*}{\partial \bar{B}_j}, \quad j \geq 1. \tag{9.35}$$

It follows from (9.32) that we have the anti-symmetries

$$\frac{\partial \mathcal{H}_j}{\partial M_k} + \frac{\partial \mathcal{H}_k}{\partial M_j} = 0, \quad k, j \geq 1. \tag{9.36}$$

In particular

$$\frac{\partial \mathcal{H}_j}{\partial M_j} = 0, \quad j \geq 1. \tag{9.37}$$

By (9.27) the latter equation also holds for $j = 0$.

Multiplying (9.32) with $\frac{\partial f}{\partial \zeta}$, multiplying (9.30) with $\frac{\partial f}{\partial M_j}$ and subtracting the results gives

$$\left(\frac{\partial f}{\partial \zeta}\frac{\partial f^*}{\partial M_j} - \frac{\partial f}{\partial M_j}\frac{\partial f^*}{\partial \zeta}\right)\frac{\partial f}{\partial M_k} = \frac{\partial f}{\partial \zeta}\frac{\partial \mathcal{H}_k}{\partial M_j} - \frac{\partial f}{\partial M_j}\frac{\partial \mathcal{H}_k}{\partial \zeta}.$$

For $j = 0$ this becomes,

$$\{f, f^*\}\frac{\partial f}{\partial M_k} = \{f, \mathcal{H}_k\},$$

hence, in view of (9.29),

$$\frac{\partial f}{\partial M_k} = \{f, \mathcal{H}_k\}.$$

Arguing similarly for $\tilde{\mathcal{H}}_k$, $\mathcal{K}_j$, $\tilde{\mathcal{K}}_j$ we summarize:

Theorem 9.1 *The dynamics of f in terms of the moments and other parameters is given by the equations*

$$\{f, f^*\} = 1,$$

$$\frac{\partial f}{\partial M_k} = \{f, \mathcal{H}_k\}, \qquad \frac{\partial f}{\partial \overline{M}_k} = \{f, \tilde{\mathcal{H}}_k\}, \quad k \geq 1,$$

$$\frac{\partial f}{\partial B_j} = \{f, \mathcal{K}_j\}, \qquad \frac{\partial f}{\partial \overline{B}_j} = \{f, \tilde{\mathcal{K}}_j\}, \quad j \geq 1.$$

In addition, we have the anti-symmetries (9.36), and similarly for the other Hamiltonians.

Besides what is stated in the theorem we have $\frac{\partial f}{\partial M_0} = \{f, \mathcal{H}_0\}$. But this equation is tautological, as follows from (9.26), (9.27).

Turning next to the structure of the Hamiltonians we have:

Theorem 9.2 *For $k \geq 1$, the Hamiltonians $\mathcal{H}_k$, $\tilde{\mathcal{H}}_k$ are rational functions of ζ of the form*

$$\mathcal{H}_k(\zeta) = \frac{b_k}{\zeta^k} + \frac{b_{k-1}}{\zeta^{k-1}} + \cdots + \frac{b_1}{\zeta} + b_0,$$

$$\tilde{\mathcal{H}}_k(\zeta) = -\bar{b}_0 - \bar{b}_1\zeta - \bar{b}_2\zeta^2 + \cdots - \bar{b}_k\zeta^k.$$

Here $b_j = -jc_j$ for $j \geq 1$, where the c_j are the same as in (8.25). The differentials with respect to ζ (only) are, in the same notation,

$$d\mathcal{H}_k(\zeta) = L_k(\zeta; f)d\zeta = \left(\frac{c_k}{\zeta^{k+1}} + \frac{c_{k-1}}{\zeta^k} + \cdots + \frac{c_1}{\zeta^2}\right) d\zeta,$$

$$d\tilde{\mathcal{H}}_k(\zeta) = -L_k^*(\zeta; f)d\zeta = -(\bar{c}_1 + \bar{c}_2\zeta + \cdots + \bar{c}_k\zeta^{k+1}) d\zeta.$$

For $k = 0$,

$$\mathcal{H}_0(\zeta) = -\log\zeta + b_0, \qquad \frac{\partial b_0}{\partial M_0} = 0,$$

$$d\mathcal{H}_0(\zeta) = -L_0(\zeta; f)d\zeta = -\frac{d\zeta}{\zeta}.$$

Proof The discussions in Example 9.1 about extensions to the Schottky double apply in principle, however with the difference that we now have to do with complex time parameters, like $t_k = M_k$. The differential $d\mathcal{H}_k$ will still be meromorphic on the double of Ω, with a pole at $z = 0$ with singular part $z^{-(k+1)}dz$, but it will no longer be symmetric on the double, with a corresponding pole on the reverse side.

Instead that pole is taken care of by the conjugate time $\bar{t}_k = \bar{M}_k$. In the formula corresponding to (9.14), the two conjugate times will correspond separately to the two terms in the right member.

For $\mathcal{H}_k$ and $d\mathcal{H}_k$ the conclusion is that they are rational with respect to ζ, precisely

$$\mathcal{H}_k(\zeta) = \left(\frac{1}{kf(\zeta)^k}\right)_{<}, \quad d\mathcal{H}_k(\zeta) = -\left(\frac{f'(\zeta)d\zeta}{f(\zeta)^{k+1}}\right)_{<} d\zeta,$$

$$\frac{\partial \mathcal{H}_k(\zeta)}{\partial \zeta} = -\left(\frac{f'(\zeta)d\zeta}{f(\zeta)^{k+1}}\right)_{<}.$$

As in (8.25), the subscript $<$ indicates singular part of a Laurent series. From the above Theorem 9.2 follows easily. □

Remark 9.1 The detailed structure of the Hamiltonians $\mathcal{H}_j$ related to the branch points remains a question for future investigation. Similarly, the additive constant b_0 in Theorem 9.2 seem to be difficult to get hold of. The coefficients c_j are determined in terms of the data for f by (8.25), and for $j \geq 1$ we have $b_j = -jc_j$. Ideally we would like to express the Hamiltonians directly in terms of $M_0, M_1, \ldots,$ $B_1, B_2, \ldots,$ besides the variable ζ, but to do so explicitly seems to be difficult.

Example 9.2 On writing (with $a_0 > 0$)

$$f(\zeta) = a_0\zeta + a_1\zeta^2 + a_2\zeta^3 + \ldots,$$

we have

$$\mathcal{H}_1(\zeta) = \frac{1}{a_0\zeta} + b_0, \quad \tilde{\mathcal{H}}_1(\zeta) = -\bar{b}_0 - \frac{\zeta}{a_0},$$

$$d\mathcal{H}_1(\zeta) = -\frac{d\zeta}{a_0\zeta^2}, \quad d\tilde{\mathcal{H}}_1(\zeta) = -\frac{d\zeta}{a_0},$$

$$\mathcal{H}_2(\zeta) = \frac{1}{2a_0^2\zeta^2} - \frac{a_1}{a_0^3\zeta} + b_0, \quad \tilde{\mathcal{H}}_2(\zeta) = -\bar{b}_0 + \frac{\bar{a}_1\zeta}{a_0^3} - \frac{\zeta^2}{2a_0^2},$$

$$d\mathcal{H}_2(\zeta) = \left(-\frac{1}{a_0^2\zeta^3} + \frac{a_1}{a_0^3\zeta^2}\right)d\zeta, \quad d\tilde{\mathcal{H}}_2(\zeta) = \left(\frac{\bar{a}_1}{a_0^3} - \frac{\zeta}{a_0^2}\right)d\zeta.$$

Chapter 10
The String Equation for Some Rational Functions

Abstract We discuss the string equation in some more depth in cases when all but finitely many moments vanish. This assumption amounts to the domain being a quadrature Riemann surface of a certain form, and it allows for a rigorous treatment of the problems involved. It applies in particular to the case of polynomial mapping functions, in which case the treatment becomes fully algebraic and involves an explicit formula for the Jacobi determinant for the map from coefficients to harmonic moments.

10.1 The String Equation on Quadrature Riemann Surfaces

One way to handle in a rigorous way the problem, mentioned in the beginning of Sect. 8.3, that the harmonic moments represent too few test functions because functions in the z-plane cannot not distinguish points on different sheets on the Riemann surface above it, is to turn to the class of quadrature Riemann surfaces. Such surfaces have been introduced and discussed in a special case in [100] (see also [46]), and we shall need them only in that special case. In principle, a quadrature Riemann surface is a Riemann surface provided with a Riemannian metric such that a finite quadrature identity holds for the corresponding area integral of integrable analytic functions on the surface.

The special case which we shall consider is that the Riemann surface is a bounded simply connected branched covering surface of the complex plane and that the quadrature identity evaluates at only one point, chosen to be the origin. When pulled back to the unit disk with a map $f \in \mathscr{O}_{\mathrm{norm}}(\overline{\mathbb{D}})$ it is then a special case of the identity (6.25) in Proposition 6.1, namely with $c_1 = \cdots = c_r = 0$, $\alpha_0 = 0$, $\ell = 0$, in the notations of that proposition.

We write this in new notations as

$$\frac{1}{2\pi \mathrm{i}} \int_{\mathbb{D}} h(\zeta) |f'(\zeta)|^2 \, d\bar{\zeta} d\zeta = \sum_{j=0}^{n} c_j h^{(j)}(0), \qquad (10.1)$$

which is to hold for analytic test functions h which are integrable with respect to the weight $|f'|^2$. The coefficients c_j are fixed complex constants (c_0 necessarily real and positive), and we assume that $c_n \neq 0$ to give the integer n a definite meaning.

By Proposition 6.1, $g = f'$ is necessarily a rational function when an identity (10.1) holds, and inspection of the proof reveals that f is actually rational itself when all the terms with line integrals in (6.25) vanish. Indeed, the function $Q(z)$ in (6.26) will vanish identically, and then it is immediate from (6.26) that the two members in that relation glue together to a meromorphic function on the Riemann sphere, which forces f to be a rational function.

We point out that quadrature Riemann surfaces as above are dense in the class of all bounded simply connected branched covering surfaces over $\mathbb{C}$. Therefore the restriction to quadrature Riemann surfaces is no severe restriction. The reason that (10.1) is a useful identity in our context is that it reduces the information of $f(\mathbb{D})$ as a multi-sheeted surface to information concentrated at one single point on it, namely $f(0)$ regarded as a point on $f(\mathbb{D})$, and near that point it does not matter that the test functions $1, z, z^2, \ldots$ cannot distinguish different sheets from each other.

Having an identity (10.1) we can easily compare the constants $(c_0, c_1, \ldots, c_n)$ with the moments $(M_0, M_1, \ldots, M_n)$. It is just to choose $h(\zeta) = f(\zeta)^k$ to obtain M_k, and this gives, for fixed f, a linear relationship mediated by a non-singular triangular matrix. Thus we have a one-to one correspondence

$$(M_0, M_1, \ldots, M_n) \leftrightarrow (c_0, c_1, \ldots, c_n).$$

The relations between the c_j and f are obtained by writing the left member of (10.1) as

$$\frac{1}{2\pi i} \int_{\partial \mathbb{D}} h(\zeta) f^*(\zeta) f'(\zeta) d\zeta = \operatorname*{Res}_{\zeta=0} h(\zeta) f^*(\zeta) f'(\zeta) d\zeta.$$

Comparing this with the right member of the same equation shows that

$$f^*(\zeta) f'(\zeta) = \sum_{k=0}^{n} \frac{k! c_k}{\zeta^{k+1}} + \text{holomorphic in } \mathbb{D}. \tag{10.2}$$

Hence the information about the moments is encoded in local information of f at the origin and infinity.

If f' has zeros $\omega_1, \ldots, \omega_m$ in $\mathbb{D}$ then (10.2) means that f^* is allowed to have poles at these points, in addition to the necessary pole of order $n + 1$ at the origin, which is implicit in (10.2) since $f'(0) \neq 0$. We may now start counting parameters. Taking into account the normalization at the origin, f has from start $(1+2m+2n)+2m$ real parameters (numerator plus denominator when writing f as a quotient). These shall be matched with the $1 + 2n$ parameters in the M_k or c_k. Next, each pole of f^* in $\mathbb{D} \setminus \{0\}$ has to be a zero of f', which give $2m$ equations for the parameters. Now there remain $2m$ free parameters, and we claim that these can be taken to be the locations of the branch points, namely

$$B_j = f(\omega_j) \quad j = 1, \ldots, m. \tag{10.3}$$

Thus we expect that f can be parametrized by the M_k and the B_j:

$$f(\zeta) = f(\zeta, M_0, \ldots, M_n, B_1, \ldots, B_m). \tag{10.4}$$

In particular $\partial f/\partial M_0$ then makes sense, with the understanding that $B_1, \ldots, B_m$, as well as $M_1, \ldots, M_n$, are kept fixed under the derivation.

Clearly the parameters M_k and B_j depend smoothly on f. This dependence can be made explicit by obvious residue formulas:

$$M_k = \frac{1}{2\pi i} \int_{\partial \mathbb{D}} f(\zeta)^k f^*(\zeta) f'(\zeta) d\zeta = \operatorname*{Res}_{\zeta=0} f(\zeta)^k f^*(\zeta) f'(\zeta) d\zeta, \tag{10.5}$$

$$B_j = \frac{1}{2\pi i} \oint_{|\zeta - \omega_j| = \varepsilon} \frac{f(\zeta) f''(\zeta)}{f'(\zeta)} d\zeta = \operatorname*{Res}_{\zeta=\omega_j} \frac{f(\zeta) f''(\zeta)}{f'(\zeta)} d\zeta. \tag{10.6}$$

We summarize part of the above discussion in Theorem 10.1 below, and then formulate the main result in this context, Theorem 10.2.

Theorem 10.1 *Let $f \in \mathcal{O}_{\mathrm{norm}}(\overline{\mathbb{D}})$, satisfying $f' \neq 0$ on $\partial \mathbb{D}$. Let $\omega_1, \ldots, \omega_m$ denote the zeros of f' in $\mathbb{D}$, these zeros assumed to be simple. Then a quadrature identity of the kind (10.1) holds, for some choice of coefficients $c_0, c_1, \ldots, c_n$ with $c_n \neq 0$, if and only if f is a rational function such that f has a pole of order $n + 1$ at infinity and possibly finite poles at the reflected points $1/\bar{\omega}_k$ of the zeros of f' in $\mathbb{D}$. This means that f is of the form*

$$f(\zeta) = \frac{a_0\zeta + a_1\zeta^2 + \cdots + a_{m+n}\zeta^{m+n+1}}{(1 - \bar{\omega}_1\zeta) \ldots (1 - \bar{\omega}_m\zeta)}. \tag{10.7}$$

Theorem 10.2 *For functions f as in Theorem 10.1, the coefficients $a_0, \ldots, a_{m+n}$ can be viewed as free parameters, subject only to $a_0 > 0$, with the roots $\omega_1, \ldots, \omega_m$ being determined by the condition $f'(\omega_k) = 0$ $(k = 1, \ldots, m)$. The map*

$$(a_0, a_1, \ldots, a_{m+n}) \mapsto (M_0, M_1, \ldots, M_n, B_1, \ldots, B_m)$$

obtained from (10.5), (10.6) (or (10.3)) can then be viewed as a change of local coordinates. In terms of the latter coordinates the partial derivative $\partial f/\partial M_0$, and hence the Poisson bracket (8.5), makes sense. And the string equation holds:

$$\{f, f^*\} = 1. \tag{10.8}$$

Proof The first statements, concerning the form of f, were proved already in the text preceding the theorem, and can in addition be viewed as a special case of Proposition 6.1.

As remarked before the proof, the parameters $M_0, M_1, \ldots, M_n, B_1, \ldots, B_m$ represent as many data as there are independent coefficients in (10.7), namely the $a_0, a_1, \ldots, a_{m+n}$, and it is implicit in previous considerations that they are indeed independent coordinates. More precisely, pure motions of branch points are regulated by the last term $R(\zeta, t)$ in (2.26), and it is seen from (2.29) that their speeds (with respect changes of M_0) are proportional to the free coefficients B_{j1} in $R(\zeta, t)$ (note that $f''(\omega_j) \neq 0$ under present assumptions).

Finally we know from Proposition 6.1 that there exists, given $f(\cdot, 0)$, an evolution $t \mapsto f(\cdot, t)$ such that $M_1, \ldots, M_n, B_1, \ldots, B_m$ remain fixed under the evolution and such that (10.8) holds. $\square$

10.2   The String Equation for Polynomials

This section follows [33] closely, but in subsequent sections we also illustrate the treatment by an example which has previously been discussed by C. Huntingford [57], and we also make some remarks on inverse problems in potential theory.

We restrict to polynomials, of a fixed degree $n + 1$:

$$f(\zeta) = \sum_{j=0}^{n} a_j \zeta^{j+1}, \quad a_0 > 0. \tag{10.9}$$

The derivative is of degree n, and we denote its coefficients by b_j:

$$g(\zeta) = f'(\zeta) = \sum_{j=0}^{n} b_j \zeta^j = \sum_{j=0}^{n} (j+1) a_j \zeta^j. \tag{10.10}$$

For polynomials f as above all moments $M_k, \bar{M}_k$ with $k > n$ vanish. For univalent (or locally univalent) functions $f \in \mathscr{O}_{\mathrm{norm}}(\overline{\mathbb{D}})$ the vanishing of all but finitely many moments is actually equivalent to being a polynomial of the form (10.9), but as is clear from Sect. 10.1, non-univalent rational functions f which are not polynomials can also have this property.

It is obvious from Definition 8.1 that the string equation cannot hold if $g = f'$ has zeros at two points which are reflections of each other with respect to the unit circle. In other words, the string equation cannot hold if g and g^* have common zeros. The main result in this section says that for polynomial maps this is the only exception: whenever g and g^* have no common zeros, the moments $M_0, M_1, \bar{M}_1, \ldots, M_n, \bar{M}_n$ serve as good local coordinates and the string equation holds. Note that these moments match the coefficients in (10.9), both representing $2n + 1$ real degrees of freedom. For such reasons the branch points will not be needed as additional parameters when restricting to functions of the form (10.9).

Next recall that the meromorphic resultant $\mathscr{R}(g, g^*)$, introduced in Sect. 3.6 (Definition 3.4), has a property related to the condition above, namely that it is nonzero if and only if g and g^* have no common zeros (these functions never have any common pole). Therefore this resultant should naturally enter the discussion. Indeed, the main result, Theorem 10.3 below, is an interplay between the Poisson bracket, the resultant and the Jacobi determinant between the moments and the coefficients of f in (10.9). The first, and major, part namely (10.11), is due to O. Kuznetsova and V. Tkachev [66, 116]. The string equation then comes out as a rather simple consequence.

Theorem 10.3 *With f a polynomial as in (10.9) and $g = f'$, the identity*

$$\frac{\partial(\bar{M}_n, \ldots \bar{M}_1, M_0, M_1, \ldots, M_n)}{\partial(\bar{a}_n, \ldots, \bar{a}_1, a_0, a_1, \ldots, a_n)} = 2a_0^{n^2+3n+1}\mathscr{R}(g, g^*) \tag{10.11}$$

holds generally. It follows that the derivative $\partial f/\partial M_0$ makes sense whenever $\mathscr{R}(g, g^*) \neq 0$, and then also the string equation

$$\{f, f^*\} = 1 \tag{10.12}$$

holds.

Proof Using Lemma 8.1 (or directly Eq. (8.24)) we shall first investigate how the moments change under a general variation of f, i.e. we let $f(\zeta) = f(\zeta, t)$ depend smoothly on a real parameter t. Thus $a_j = a_j(t)$, $M_k = M_k(t)$. For the Laurent series of any function $h(\zeta) = \sum_i c_i \zeta^i$ we denote by $\operatorname{coeff}_i(h)$ the coefficient of ζ^i:

$$\operatorname{coeff}_i(h) = c_i = \frac{1}{2\pi i} \oint_{|\zeta|=1} \frac{h(\zeta)d\zeta}{\zeta^{i+1}}.$$

By Lemma 8.1 we then have, for $k \geq 0$,

$$\frac{d}{dt}M_k = \frac{1}{2\pi i}\frac{d}{dt}\int_{\mathbb{D}} f(\zeta, t)^k |f'(\zeta, t)|^2 d\bar{\zeta}d\zeta = \frac{1}{2\pi i}\int_{\partial\mathbb{D}} f^k \{f, f^*\}_t \frac{d\zeta}{\zeta}$$

$$= \operatorname{coeff}_0(f^k \{f, f^*\}_t) = \sum_{i=0}^{n} \operatorname{coeff}_{+i}(f^k) \cdot \operatorname{coeff}_{-i}(\{f, f^*\}_t).$$

Note that $f(\zeta)^k$ contains only positive powers of ζ and that $\{f, f^*\}_t$ (recall (2.9)) contains powers with exponents in the interval $-n \leq i \leq n$ only.

In view of (10.9) the matrix

$$v_{ki} = \operatorname{coeff}_{+i}(f^k) \quad (0 \leq k, i \leq n)$$

is upper triangular, i.e. $v_{ki} = 0$ for $0 \leq i < k$, with diagonal elements being powers of a_0:

$$v_{kk} = a_0^k.$$

Next we shall find the coefficients of the Poisson bracket. These will involve the coefficients b_k and $\dot{a}_j$, but also their complex conjugates. For a streamlined treatment it is convenient to introduce coefficients with negative indices to represent the complex conjugated quantities. The same for the moments. Thus we define, for the purpose of this proof and the forthcoming Example 10.1,

$$M_{-k} = \bar{M}_k, \quad a_{-k} = \bar{a}_k, \quad b_{-k} = \bar{b}_k \quad (k > 0). \tag{10.13}$$

Turning points are the real quantities M_0 and $a_0 = b_0$.

In this notation the expansion of the Poisson bracket becomes

$$\begin{aligned}
\{f, f^*\}_t &= f'(\zeta) \cdot \zeta \dot{f}^*(\zeta) + f'^*(\zeta) \cdot \zeta^{-1} \dot{f}(\zeta) \\
&= \sum_{\ell, j \geq 0} b_\ell \dot{\bar{a}}_j \zeta^{\ell-j} + \sum_{\ell, j \leq 0} \bar{b}_\ell \dot{a}_j \zeta^{j-\ell} \\
&= \sum_{\ell \geq 0,\, j \leq 0} b_\ell \dot{a}_j \zeta^{\ell+j} + \sum_{\ell \leq 0,\, j \geq 0} b_\ell \dot{a}_j \zeta^{\ell+j} \\
&= b_0 \dot{a}_0 + \sum_{\ell \cdot j \leq 0} b_\ell \dot{a}_j \zeta^{\ell+j} \\
&= b_0 \dot{a}_0 + \sum_i \left(\sum_{\ell \cdot j \leq 0,\, \ell+j=-i} b_\ell \dot{a}_j \right) \zeta^{-i}. \tag{10.14}
\end{aligned}$$

The last summation runs over pairs of indices (ℓ, j) having opposite sign (or at least one of them being zero) and adding up to $-i$. We presently need only to consider the case $i \geq 0$. Eliminating ℓ and letting j run over those values for which $\ell \cdot j \leq 0$ we therefore get

$$\text{coeff}_{-i}(\{f, f^*\}_t) = b_0\, \dot{a}_0\, \delta_{i0} + \sum_{j \leq -i} b_{-(i+j)}\, \dot{a}_j + \sum_{j \geq 0} b_{-(i+j)}\, \dot{a}_j.$$

Here δ_{ij} denotes the Kronecker delta. Setting, for $i \geq 0$,

$$u_{ij} = \begin{cases} b_{-(i+j)} + b_0 \delta_{i0} \delta_{0j}, & \text{if } -n \leq j \leq -i \text{ or } 0 \leq j \leq n, \\ 0 & \text{in remaining cases} \end{cases}$$

we thus have

$$\operatorname{coeff}_{-i}(\{f, f^*\}_t) = \sum_{j=0}^{n} u_{ij} \dot{a}_j. \tag{10.15}$$

Turning to the complex conjugated moments we have, with $k < 0$,

$$\dot{M}_k = \dot{\bar{M}}_{-k} = \sum_{i \le 0} \overline{\operatorname{coeff}_{-i}(f^{-k})} \cdot \overline{\operatorname{coeff}_{+i}(\{f, f^*\}_t)}.$$

Set, for $k < 0, i \le 0$,

$$v_{ki} = \overline{\operatorname{coeff}_{-i}(f^{-k})}.$$

Then $v_{ki} = 0$ when $k < i \le 0$, and $v_{kk} = a_0^{-k}$. To achieve the counterpart of (10.15) we define, for $i \le 0$,

$$u_{ij} = \begin{cases} b_{-(i+j)} + b_0 \delta_{i0} \delta_{0j}, & \text{if } -n \le j \le 0 \text{ or } -i \le j \le n, \\ 0 & \text{in remaining cases.} \end{cases}$$

This gives, with $i \le 0$,

$$\overline{\operatorname{coeff}_{+i}(\{f, f^*\}_t)} = \sum_{j=0}^{n} u_{ij} \dot{a}_j.$$

As a summary we have

$$\dot{M}_k = \sum_{-n \le i, j \le n} v_{ki} u_{ij} \dot{a}_j, \quad -n \le k \le n, \tag{10.16}$$

where

$v_{ki} = \operatorname{coeff}_{+i}(f^k)$	when $0 \le k \le i$,
$v_{ki} = \overline{\operatorname{coeff}_{-i}(f^{-k})}$	when $i \le k < 0$,
$v_{ki} = 0$	in remaining cases,
$u_{ij} = b_{-(i+j)} + b_0 \delta_{i0} \delta_{0j}$	in index intervals made explicit above,
$u_{ij} = 0$	in remaining cases.

We see that the full matrix $V = (v_{ki})$ is triangular in each of the two blocks along the main diagonal and vanishes completely in the two remaining blocks. Therefore, its determinant is simply the product of the diagonal elements. More precisely this becomes

$$\det V = a_0^{n(n+1)}. \tag{10.17}$$

The matrix $U = (u_{ij})$ represents the linear dependence of the bracket $\{f, f^*\}_t$ on $g = f'$ and $g^* = f'^*$, and it acts on the column vector with components $\dot{a}_j$, then representing the linear dependence on $\dot{f}$ and $\dot{f}^*$. The computation started at (10.14) can thus be finalized as

$$\{f, f^*\}_t = \sum_{-n \leq i,j \leq n} u_{ij} \dot{a}_j \zeta^{-i}, \quad -n \leq k \leq n. \tag{10.18}$$

Returning to (10.16), this equation says that the matrix of partial derivatives $\partial M_k / \partial a_j$ equals the matrix product VU, in particular that

$$\frac{\partial(M_{-n}, \ldots M_{-1}, M_0, M_1, \ldots, M_n)}{\partial(a_{-n}, \ldots, a_{-1}, a_0, a_1, \ldots, a_n)} = \det V \cdot \det U.$$

The first determinant was already computed above, see (10.17). It remains to connect $\det U$ to the meromorphic resultant $\mathscr{R}(g, g^*)$.

For any kind of evolution, $\{f, f^*\}_t$ vanishes whenever g and g^* have a common zero. The meromorphic resultant $\mathscr{R}(g, g^*)$ is a complex number which has the same non-vanishing property, and it is in a certain sense minimal in that respect. From this one may expect that the determinant of U is simply a multiple of the resultant. Taking homogenieties into account indicates that the constant of proportionality should be b_0^{2n+1}, times possibly some numerical factor. The precise formula in fact turns out to be

$$\det U = 2b_0^{2n+1} \mathscr{R}(g, g^*). \tag{10.19}$$

One way to prove it is to connect U to the Sylvester matrix S associated to the classical polynomial resultant (see [118]) for the polynomials $g(\zeta)$ and $\zeta^n g^*(\zeta)$. This Sylvester matrix is of size $2n \times 2n$. By some operations with rows and columns (the details are given in [44], and will in addition be illustrated in the example below) one finds that

$$\det U = 2b_0 \det S.$$

From this (10.19) follows.

Now, the string equation is an assertion about a special evolution. The string equation says that $\{f, f^*\}_t = 1$ for that kind of evolution for which $\partial/\partial t$ means $\partial/\partial M_0$, in other words in the case that $\dot{M}_0 = 1$ and $\dot{M}_k = 0$ for $k \neq 0$. By what has already been proved, a unique such evolution exists with f kept on the form (10.9), as long as $\mathscr{R}(g, g^*) \neq 0$.

Inserting $\dot{M}_k = \delta_{k0}$ into (10.16) gives

$$\sum_{-n \le i,j \le n} v_{ki} u_{ij} \dot{a}_j = \delta_{k0}, \quad -n \le k \le n. \tag{10.20}$$

It is easy to see from the structure of the matrix $V = (v_{ki})$ that the 0:th column of the inverse matrix V^{-1}, which is sorted out when V^{-1} is applied to the right member in (10.20), is simply the unit vector with components δ_{k0}. Therefore (10.20) is equivalent to

$$\sum_{-n \le j \le n} u_{ij} \dot{a}_j = \delta_{i0}, \quad -n \le i \le n. \tag{10.21}$$

Inserting this into (10.18) shows that the string equation indeed holds.   □

Example 10.1 To illustrate the above proof, and the general theory, we compute everything explicitly when $n = 2$, i.e. with

$$f(\zeta) = a_0 \zeta + a_1 \zeta^2 + a_2 \zeta^3.$$

We shall keep the convention (10.13) in this example. Thus

$$f'(\zeta) = g(\zeta) = b_0 + b_1 \zeta + b_2 \zeta^2 = a_0 + 2a_1 \zeta + 3a_2 \zeta^2,$$
$$f^*(\zeta) = a_0 \zeta^{-1} + a_{-1} \zeta^{-2} + a_{-2} \zeta^{-3},$$

for example. When the Eq. (10.16) is written as a matrix equation it becomes (with zeros represented by blanks)

$$\begin{pmatrix} \dot{M}_{-2} \\ \dot{M}_{-1} \\ \dot{M}_0 \\ \dot{M}_1 \\ \dot{M}_2 \end{pmatrix} = \begin{pmatrix} a_0^2 & & & & \\ a_{-1} & a_0 & & & \\ & & 1 & & \\ & & & a_0 & a_1 \\ & & & & a_0^2 \end{pmatrix} \begin{pmatrix} & & b_2 & & b_0 & \\ & b_2 & b_1 & & b_0 & b_{-1} \\ b_2 & b_1 & 2b_0 & b_{-1} & b_{-2} \\ b_1 & b_0 & b_{-1} & b_{-2} & \\ b_0 & & b_{-2} & & \end{pmatrix} \begin{pmatrix} \dot{a}_{-2} \\ \dot{a}_{-1} \\ \dot{a}_0 \\ \dot{a}_1 \\ \dot{a}_2 \end{pmatrix}. \tag{10.22}$$

Denoting the two 5×5 matrices by V and U respectively it follows that the corresponding Jacobi determinant is

$$\frac{\partial(M_{-2}, M_{-1}, M_0, M_1, M_2)}{\partial(a_{-2}, a_{-1}, a_0, a_1, a_2)} = \det V \cdot \det U = a_0^6 \cdot \det U.$$

Here U can essentially be identified with the Sylvester matrix for the resultant $\mathcal{R}(g, g^*)$. To be precise,

$$\det U = 2b_0 \det S, \tag{10.23}$$

where S is the classical Sylvester matrix associated to the two polynomials $g(\zeta)$ and $\zeta^2 g^*(\zeta)$, namely

$$
S = \begin{pmatrix}
b_2 & & b_0 & \\
b_2 & b_1 & b_0 & b_{-1} \\
b_1 & b_0 & b_{-1} & b_{-2} \\
b_0 & & b_{-2} &
\end{pmatrix}.
$$

As promised in the proof above, we shall explain in this example the column operations on U leading from U to S, and thereby proving (10.23) in the case $n = 2$ (the general case is similar). Recall that U is the second matrix in the right member of (10.22). Let U_{-2}, U_{-1}, U_0, U_1, U_2 denote its columns. We make the following changes of U_0:

$$
U_0 \mapsto \frac{1}{2} U_0 - \frac{1}{2b_0}(b_{-2}U_{-2} + b_{-1}U_{-1} - b_1 U_1 - b_2 U_2)
$$

The first term makes the determinant become half as big as it was before, and the other terms do not affect the determinant at all. The new matrix is the 5×5 matrix

$$
\begin{pmatrix}
b_2 & & b_0 & & \\
b_2 & b_1 & b_0 & b_{-1} & \\
b_2 & b_1 & b_0 & b_{-1} & b_{-2} \\
b_1 & b_0 & & b_{-2} & \\
b_0 & & & &
\end{pmatrix}
$$

which has b_0 in the lower left corner, with the complementary 4×4 block being exactly S. From this (10.23) follows.

The string equation (10.12) becomes, in terms of coefficients and with $\dot{a}_j$ interpreted as $\partial a_j / \partial M_0$, the linear equation

$$
\begin{pmatrix}
a_0^2 & & & & \\
a_{-1} & a_0 & & & \\
& & 1 & & \\
& & & a_0 & a_1 \\
& & & & a_0^2
\end{pmatrix}
\begin{pmatrix}
b_2 & & b_0 & & \\
b_2 & b_1 & b_0 & b_{-1} & \\
b_2 & b_1 & 2b_0 & b_{-1} & b_{-2} \\
b_1 & b_0 & b_{-1} & b_{-2} & \\
b_0 & & b_{-2} & &
\end{pmatrix}
\begin{pmatrix}
\dot{a}_{-2} \\
\dot{a}_{-1} \\
\dot{a}_0 \\
\dot{a}_1 \\
\dot{a}_2
\end{pmatrix}
=
\begin{pmatrix}
0 \\
0 \\
1 \\
0 \\
0
\end{pmatrix}.
$$

Indeed, in view of (10.22) this equation characterizes the $\dot{a}_i$ as those belonging to an evolution such that $\dot{M}_0 = 1$, $\dot{M}_k = 0$ for $k \neq 0$.

10.3   Evolution of a Third Degree Polynomial with Real Coefficients

Several authors have discussed Hele-Shaw evolution of a polynomial domain of degree three. C. Huntingford [57] showed they can develop resolvable cusps. S. Howison [56] had earlier given similar examples in the case of opposite geometry (suction at infinity), and also (with coauthors) discussed cusp development in general, see [55] for example. There is in fact an extensive literature on cusp development for Hele-Shaw evolutions. We refer to [43] for more references, and [35] for the Huntingford example in particular. Here we shall look at the latter example from the point of view the resultant $\mathscr{R}(g, g^*)$ discussed in Example 10.1. Further analysis of the Hele-Shaw evolution of third degree polynomials, beyond what is discussed below, is given [73].

In general we compute the determinant for the matrix U in (10.22) as

$$
\det U = \det \begin{pmatrix} & b_2 & & b_0 & & \\ & b_2 & b_1 & b_0 & b_{-1} & \\ & b_2 & b_1 & 2b_0 & b_{-1} & b_{-2} \\ & b_1 & b_0 & b_{-1} & b_{-2} & \\ & b_0 & & b_{-2} & & \end{pmatrix}
$$

$$
= 2b_0^5 - 2b_0^3|b_1|^2 - 4b_0^3|b_2|^2 + 2b_0^2(\bar{b}_1^2 b_2 + b_1^2 \bar{b}_2) - 2b_0|b_1|^2|b_2|^2 + 2b_0|b_2|^4.
$$

In the sequel we assume that the a_j, and hence b_j, are real. Then the above determinant factorizes as

$$
\det U = 2b_0(b_0 - b_2)^2(b_0 + b_1 + b_2)(b_0 - b_1 + b_2),
$$

which gives (in terms of the a_j and using (10.19))

$$
\mathscr{R}(g, g^*) = \frac{a_0(a_0 - 3a_2)^2(a_0 + 2a_1 + 3a_2)(a_0 - 2a_1 + 3a_2)}{a_0^5}. \tag{10.24}
$$

To get a better picture we may, as in [57], pass to scaled coefficients A_1, A_2 defined by

$$
A_j = \frac{a_j}{a_0} \quad (j = 0, 1, 2), \tag{10.25}
$$

thus with $A_0 = 1$. Then

$$
\mathscr{R}(g, g^*) = (1 - 3A_2)^2(1 + 2A_1 + 3A_2)(1 - 2A_1 + 3A_2). \tag{10.26}
$$

We see that the resultant vanishes on the three straight lines

$$L_1: \quad 1 - 3A_2 = 0,$$
$$L_2: \quad 1 + 2A_1 + 3A_2 = 0,$$
$$L_3: \quad 1 - 2A_1 + 3A_2 = 0$$

in the (A_1, A_2)-plane. The complement $\mathbb{C} \setminus (L_0 \cup L_1 \cup L_2)$ has seven components one of which (the bounded one) is the triangle of local univalence.

We consider now the evolution of

$$f(\zeta, t) = a_0(t)\zeta + a_1(t)\zeta^2 + a_2(t)\zeta^3$$

under the dynamics of (2.1), with $q(t)$ chosen so that we are in the coefficient normalization case

$$a_0(t) = e^t. \tag{10.27}$$

Thus $\dot{a}_0 = a_0$ and $a_0(0) = 1$, and a point in the (A_1, A_2)-plane can be thought of as the starting values (a_1, a_2) for a solution trajectory in the (a_0, a_1, a_2)-space. In the (A_1, A_2)-plane one then sees only the change of shape of the domain represented by f. No point in the (A_1, A_2)-plane is attained twice for a given solution, the only exception being the "trivial" solution $f(\zeta, t) = e^t\zeta$, which becomes stationary in the (A_1, A_2)-plane.

The three nonzero moments are

$$M_0 = a_0^2 + 2a_1^2 + 3a_2^2 = a_0^2 (1 + 2A_1^2 + 3A_2^2),$$
$$M_1 = a_0^2 a_1 + 3a_0 a_1 a_2 = a_0^3 A_1(1 + 3A_2),$$
$$M_2 = a_0^3 a_2 = a_0^4 A_2,$$

where only M_0 depends on t. These equations can easily be solved for a_1, a_2 (or A_1, A_2) in terms of a_0, M_1, M_2, giving (with (10.27))

$$f(\zeta, t) = e^t\zeta + \frac{M_1\zeta^2}{e^{2t} + 3e^{-2t}M_2} + e^{-3t}M_2\zeta^3. \tag{10.28}$$

Thus

$$A_1 = \frac{M_1 e^t}{e^{4t} + 3M_2}, \quad A_2 = M_2 e^{-4t}.$$

Eliminating t between these equations gives the trajectories in the (A_1, A_2)-plane
on the form

$$A_1 = M_1 M_2^{-3/4} \cdot \frac{A_2^{3/4}}{1 + 3A_2}, \tag{10.29}$$

when restricting to the range $0 < A_2 < \frac{1}{3}$.

Next, straight-forward computations in combination with (10.26) and (2.5) give

$$q(t) = \frac{1}{2} \dot{M}_0(t) = \frac{a_0^2 (1 - 3A_2)(1 + 2A_1 + 3A_2)(1 - 2A_1 + 3A_2)}{1 + 3A_2}$$

$$= \frac{a_0^2 \mathscr{R}(g, g^*)}{(1 + 3A_2)(1 - 3A_2)}. \tag{10.30}$$

Hence the sign of $q(t)$ changes whenever one of the lines L_j is crossed, and also
when the line defined by $1 + 3A_2 = 0$ is crossed.

The coefficient range within which f is locally univalent is the triangle in the
(A_1, A_2)-plane delimited by the lines L_1, L_2, L_3, which have corners at $(-1, 1/3)$,
$(1, 1/3)$ and $(0, -1/3)$. The range of univalence is only slightly smaller: close to
the corners $(\pm 1, 1/3)$ the non-horizontal sides of the triangle are to be replaced by
the nearby segments of the ellipse

$$E: \quad A_1^2 + 4(A_2 - \frac{1}{2})^2 = 1,$$

with $\frac{1}{5} \le A_2 \le \frac{1}{3}$; see [10, 13, 57] for details and Fig. 10.1 for illustration. The
limiting values of A_2 correspond to the points $(A_1, A_2) = (\pm\frac{4}{5}, \frac{1}{5})$, at which the
ellipse is tangent to the nonhorizontal lines L_2 and L_3 of the triangle of local
univalence, and $(A_1, A_2) = (\pm\frac{2\sqrt{2}}{3}, \frac{1}{3})$, at which the ellipse intersects the line L_1.
At each of the latter points, the domain $\Omega = f(\mathbb{D})$ has two cusps and in addition a
double point on its boundary. At each of the points $(A_1, A_2) = (\pm\frac{4}{5}, \frac{1}{5})$, Ω has a
5/2-power cusp.

The Huntingford trajectory is the unique solution which passes through one of
the latter points, say $(\frac{4}{5}, \frac{1}{5})$. It hits the horizontal side L_1 at $(A_1, A_2) = (\frac{16}{15}\sqrt[4]{\frac{3}{5}}, \frac{1}{3})$,
which lies to the left of $(\frac{2\sqrt{2}}{3}, \frac{1}{3})$, hence in the region of univalence. Compare
Fig. 10.2. The conserved moments are $M_1 = \frac{32}{25}$, $M_2 = \frac{1}{5}$, while $M_0(t)$ increases
with t, however with $\dot{M}_0(0) = 0$.

As is seen from (10.28), the function $f(\zeta, t)$ is defined for all $-\infty < t < \infty$
when $M_2 \ge 0$, and it satisfies the Polubarinova-Galin equation (2.1) for all these
t. The same is true for all values of M_2, provided $M_1 = 0$. When $M_2 < 0$ and
$M_1 \ne 0$, then $a_2(t) \to \pm\infty$ as $t \to \frac{1}{4}\log(-3M_2)$. It is still true that through each
point (a_0, a_1, a_2) with $a_0 \ne 0$ in coefficient space there passes exactly one solution
trajectory. On the other hand, the Löwner-Kufarev equation (2.10) holds only within

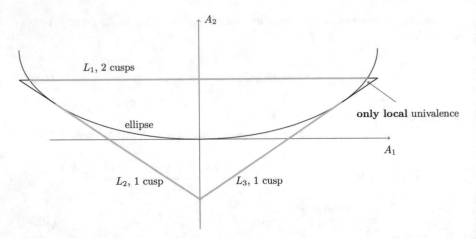

Fig. 10.1 Coefficient regions for local and global univalence in Sect. 10.3. The triangle bounded by green lines, complemented by black lines in the corners, represents region of local univalence. The region bounded by green lines, complemented by red elliptic arcs in the corners, represents region of global univalence

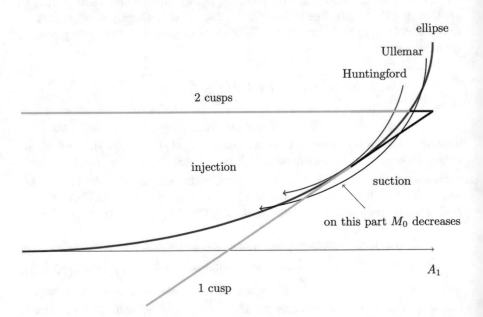

Fig. 10.2 A closer look at the Huntingford trajectory together with a nearby Hele-Shaw trajectory related to the counter example of Ullemar, as described in the text (Schematic picture)

the range of local univalence, because outside that range there will be branch points which move.

10.4 An Example by Ullemar

This example does not directly refer to the Hele-Shaw problem, but we wish to mention that C. Ullemar [117] has given an example of two different polynomials of degree three, one univalent and one locally univalent, having exactly the same moments M_k. The coefficients of the polynomials are (to three decimals)

$$\begin{cases} a_0 = 6.229 \\ a_1 = 4.598 \qquad \text{(the univalent polynomial)} \\ a_2 = 1.000 \end{cases}$$

hence $(A_1, A_2) = (0.739, 0.161)$, and

$$\begin{cases} a_0 = 5.214 \\ a_1 = 4.907 \qquad \text{(the locally univalent polynomial)} \\ a_2 = 1.705 \end{cases}$$

hence $(A_1, A_2) = (0.941, 0.326)$. The nonzero moments are

$$\begin{cases} M_0 = 84.071 \\ M_1 = 264.277 \\ M_2 = 241.647. \end{cases}$$

The factor $M_1 M_2^{-3/4}$ appearing in (10.29) is close to that one for the Huntingford trajectory, hence the two trajectories are close to each other in the (A_1, A_2)-plane, as illustrated in Fig. 10.2.

As far as we know, nobody has been able to give an example of two univalent polynomials having the same moments. On the other hand, we see no reason why such polynomials should not exist. This uniqueness question is related to a classical inverse problem in potential theory. See [4, 25, 128], for reviews and some recent results.

Ullemar's example can be well understood in terms of the Huntingford example. In fact, the two points lie on the same Hele-Shaw trajectory. Using the notations from the previous subsection, this trajectory starts at some point (A_1, A_2) with $A_2 = \frac{1}{3}$ and $\frac{16}{15}\sqrt[4]{\frac{3}{5}} < A_1 < 1$, hence in the nonunivalent (but still locally univalent) region. For some value of A_2 in the interval $\frac{1}{5} < A_2 < \frac{1}{3}$ it crosses the line L_3

and leaves the triangle of local univalence. Then, for some $A_2 < \frac{1}{5}$, it crosses the line again and returns to the triangle. The moments M_1 and M_2 are preserved all the time, while M_0 first increases, then decreases (when outside the triangle), and finally increases again. It follows that there must be pairs of points on the trajectory, and inside the triangle of local univalence, with the same value also of M_0.

By using partial balayage one can produce many examples of this kind, that is pairs of simply connected domains, one "schlicht" (i.e. single-sheeted), one non-schlicht but without branch points, such that they have the same harmonic moments. One simply starts with some $f \in \mathcal{O}_{\text{locu}}(\overline{\mathbb{D}})$ which is not univalent. Then one sweeps the counting function ν_f to density one, say

$$\text{Bal}\,(\nu_f m, m) = \chi_D\, m.$$

Compare Example 3.1. The domain D will have the same harmonic moments as the multisheeted domain $f(\mathbb{D})$, and it is easy to arrange matters so that D is simply connected.

Ideas of sweeping excessive parts of the counting functions, as above, were used already in the 1970s by M. Sakai in order to prove estimates of for capacity functions and related conformally invariant metrics, and thereby make progress on open problems stated in [106]. In fact, it seems that Sakai invented methods for constructing quadrature domains for subharmonic functions exactly to this purpose. See Chapter III in [97].

Glossary

- $\mathbb{D}(a, R) = \{\zeta \in \mathbb{C} : |\zeta - a| < R\}$.
- $\mathbb{D} = \mathbb{D}(0, 1)$.
- $\mathbb{P} = \mathbb{C} \cup \{\infty\}$ = the Riemann sphere.
- $i = \sqrt{-1}$ (other meanings of the same letter are written with the usual font i).
- $\overline{\Omega} = \operatorname{clos} \Omega$, the closure of a set Ω.
- $\Omega^e = \mathbb{C} \setminus \overline{\Omega}$, the exterior of a domain Ω in $\mathbb{C}$ (or in $\mathbb{P}$ or in some other ambient space, depending on context).
- $\operatorname{int} K$, the interior of a set K.
- χ_E: the characteristic function of a set E.
- $dm = dm(z) = dx \wedge dy = \frac{1}{2i} d\bar{z} \wedge dz$ ($z = x + iy$), area measure in the z plane.
- $\zeta^* = 1/\bar{\zeta}$, when $\zeta \in \mathbb{C}$ considered as a point.
- $f^*(\zeta) = \overline{f(1/\bar{\zeta})}$. (There is a slight ambiguity compared to previous definition.)
- $\dot{f}(\zeta, t) = \frac{\partial}{\partial t} f(\zeta, t)$, $f'(\zeta, t) = \frac{\partial}{\partial \zeta} f(\zeta, t)$.
- $g(\zeta, t) = f'(\zeta, t)$ (common notation when f is mapping function).
- With $E \subset \mathbb{C}$ any set which contains the origin,

$$\mathcal{O}(E) = \{f : f \text{ is analytic in some neighborhood of } E\},$$

$$\mathcal{O}_{\mathrm{norm}}(E) = \{f \in \mathcal{O}(E) : f(0) = 0, f'(0) > 0\},$$

$$\mathcal{O}_{\mathrm{locu}}(E) = \{f \in \mathcal{O}_{\mathrm{norm}}(E) : f' \neq 0 \text{ on } E\},$$

$$\mathcal{O}_{\mathrm{univ}}(E) = \{f \in \mathcal{O}_{\mathrm{locu}}(E) : f \text{ is univalent (one-to-one) on } E\}.$$

- $\mathcal{O}$ also denotes "ordo", i.e. order of vanishing. Example: $\mathcal{O}(t^n)$ = any function which tends to zero as least as fast as t^n when $t \to 0$ ($n \geq 0$).
- card = 'number of elements in'.
- $SL^1(\Omega, \lambda)$ denotes the set of subharmonic functions in Ω which are integrable with respect to a measure λ. See (3.6).
- ν_f: counting function, see Definition 2.1.

© The Author(s), under exclusive license to Springer Nature Switzerland AG 2021
B. Gustafsson, Y.-L. Lin, *Laplacian Growth on Branched Riemann Surfaces*,
Lecture Notes in Mathematics 2287, https://doi.org/10.1007/978-3-030-69863-8

- Bal (μ, λ): partial balayage, see Definition 3.2.
- supp μ: the closed support of a measure, or distribution, μ.
- U_μ: the logarithmic potential of a measure μ. Normalization: $-\Delta U_\mu = \mu$.
- $C_\mu = \hat{\mu} = -4\frac{\partial U_\mu}{\partial z}$: the Cauchy transform of μ. Normalization: $\frac{\partial C_\mu}{\partial \bar{z}} = \mu$.
- $G_\Omega(z, a) = -\log|z - a| + \text{harmonic}$: the Green function of a domain Ω.
- $E_\rho(z, w)$: the exponential transform with weight ρ, see Definition 3.3.
- $\mathscr{R}(g, h)$: the resultant of two meromorphic function on a compact Riemann surface, see Definition 3.4.
- $\mathscr{E}_{g,h}(z, w)$: the elimination function for two meromorphic function on a compact Riemann surface, see Definition 3.5.
- $i(\mathbf{V})$: interior multiplication by a vector field $\mathbf{V}$.
- $L_{\mathbf{V}}$: the Lie derivative with respect to a vector field $\mathbf{V}$.
- $\langle \mu, \varphi \rangle$: the action of a distribution on a test function, and more generally the action a linear functional on an element in the primary space.

References

1. A. Abanov, M. Mineev-Weinstein, A. Zabrodin, Multi-cut solutions of Laplacian growth. Phys. D **238**(17), 1787–1796 (2009)
2. D. Aharonov, H.S. Shapiro, Domains on which analytic functions satisfy quadrature identities. J. Anal. Math. **30**, 39–73 (1976)
3. O. Alekseev, M. Mineev-Weinstein, Theory of stochastic Laplacian growth. J. Stat. Phys. **168**(1), 68–91 (2017)
4. Y. Ameur, M. Helmer, F. Tellander, On the uniqueness problem for quadrature domains. Preprint, Lund University, Lund, 2020
5. V.I. Arnold, *Mathematical Methods of Classical Mechanics* (Springer, New York, 1978). Translated from the Russian by K. Vogtmann and A. Weinstein, Graduate Texts in Mathematics, vol. 60
6. C. Baiocchi, Su un problema di frontiera libera connesso a questioni di idraulica. Ann. Mat. Pura Appl. (4) **92**, 107–127 (1972)
7. C. Baiocchi, Free boundary problems in the theory of fluid flow through porous media, in *Proc. Intern. Congress Math.*, Vancouver (1974), vol. II (Canad. Math. Congres, Montreal, Que., 1975), pp. 237–243
8. C. Baiocchi, A. Capelo, *Variational and Quasivariational Inequalities*. A Wiley-Interscience Publication (Wiley, New York, 1984). Applications to free boundary problems, Translated from the Italian by Lakshmi Jayakar
9. F. Bracci, M.D. Contreras, S. Díaz-Madrigal, A. Vasil'ev, Classical and stochastic Löwner-Kufarev equations, in *Harmonic and Complex Analysis and Its Applications*. Trends Math. (Birkhäuser/Springer, Cham, 2014), pp. 39–134
10. D.A. Brannan, Coefficient regions for univalent polynomials of small degree. Mathematika **14**, 165–169 (1967)
11. R. Brime, Computation of weighted Hele-Shaw flows. Master's thesis 1999:E18, Lund University, Centre for Mathematical Sciences, Mathematics, Lund, 1999
12. L.A. Caffarelli, A remark on the Hausdorff measure of a free boundary, and the convergence of coincidence sets. Boll. Un. Mat. Ital. A (5) **18**(1), 109–113 (1981)
13. V.F. Cowling, W.C. Royster, Domains of variability for univalent polynomials. Proc. Am. Math. Soc. **19**, 767–772 (1968)
14. D. Crowdy, Quadrature domains and fluid dynamics, in *Quadrature Domains and Their applications*, vol. 156 of *Oper. Theory Adv. Appl.* (Birkhäuser, Basel, 2005), pp. 113–129
15. P.J. Davis, *The Schwarz Function and Its Applications* (The Mathematical Association of America, Buffalo, NY, 1974) The Carus Mathematical Monographs, No. 17

© The Author(s), under exclusive license to Springer Nature Switzerland AG 2021   147
B. Gustafsson, Y.-L. Lin, *Laplacian Growth on Branched Riemann Surfaces*,
Lecture Notes in Mathematics 2287, https://doi.org/10.1007/978-3-030-69863-8

16. P. Diaconis, W. Fulton, A growth model, a game, an algebra, Lagrange inversion, and characteristic classes. Rend. Sem. Mat. Univ. Politec. Torino **49**(1), 95–119 (1991/1993). Commutative algebra and algebraic geometry, II (Italian) (Turin, 1990)

17. B. Dubrovin, Hamiltonian formalism of Whitham-type hierarchies and topological Landau-Ginsburg models. Commun. Math. Phys. **145**(1), 195–207 (1992)

18. C.M. Elliott, V. Janovský, A variational inequality approach to Hele-Shaw flow with a moving boundary. Proc. Roy. Soc. Edinburgh Sect. A **88**(1–2), 93–107 (1981)

19. J. Escher, G. Simonett, Classical solutions of multidimensional Hele-Shaw models. SIAM J. Math. Anal. **28**(5), 1028–1047 (1997)

20. H.M. Farkas, I. Kra, *Riemann Surfaces*, vol. 71 of *Graduate Texts in Mathematics*, 2nd edn. (Springer, New York, 1992)

21. H. Federer, *Geometric Measure Theory*. Die Grundlehren der mathematischen Wissenschaften, Band 153 (Springer, New York, 1969)

22. T. Frankel, *The Geometry of Physics*, 3rd edn. (Cambridge University Press, Cambridge, 2012). An Introduction

23. A. Friedman, *Variational Principles and Free-Boundary Problems*. Pure and Applied Mathematics (Wiley, New York, 1982). A Wiley-Interscience Publication

24. L.A. Galin, Unsteady filtration with a free surface. C. R. (Doklady) Acad. Sci. USSR (N.S.) **47**, 246–249 (1945)

25. S. Gardiner, T. Sjodin, Convexity and the exterior inverse problem of potential theory. Proc. Am. Math. Soc. **136**(5), 1699–1703 (2008)

26. S. Gardiner, T. Sjödin, Partial balayage and the exterior inverse problem of potential theory, in *Potential Theory and Stochastics in Albac*, vol. 11 of *Theta Ser. Adv. Math.* (Theta, Bucharest, 2009), pp. 111–123

27. P. Griffiths, J. Harris, *Principles of Algebraic Geometry*. Wiley-Interscience (Wiley, New York, 1978). Pure and Applied Mathematics

28. B. Gustafsson, On a differential equation arising in a Hele-Shaw flow moving boundary problem. Ark. Mat. **22**(2), 251–268 (1984)

29. B. Gustafsson, Applications of variational inequalities to a moving boundary problem for Hele-Shaw flows. SIAM J. Math. Anal. **16**(2), 279–300 (1985)

30. B. Gustafsson. On quadrature domains and an inverse problem in potential theory. J. Anal. Math. **55**, 172–216 (1990)

31. B. Gustafsson, Lectures on balayage, in *Clifford Algebras and Potential Theory*, vol. 7 of *Univ. Joensuu Dept. Math. Rep. Ser.* (Univ. Joensuu, Joensuu, 2004), pp. 17–63

32. B. Gustafsson, Exponential transforms, resultants and moments, in *Harmonic and Complex Analysis and Its Applications*. Trends Math. (Birkhäuser/Springer, Cham, 2014), pp. 287–323

33. B. Gustafsson, The string equation for polynomials. Anal. Math. Phys. **8**, 637–653 (2018)

34. B. Gustafsson, The string equation for some rational functions, in *Analysis as a Life*. Trends Math. (Birkhäuser/Springer, Cham, 2019), pp. 213–235

35. B. Gustafsson, Y.-L. Lin, On the dynamics of roots and poles for solutions of the Polubarinova-Galin equation. Ann. Acad. Sci. Fenn. Math. **38**(1), 259–286 (2013)

36. B. Gustafsson, D. Prokhorov, A. Vasil'ev, Infinite lifetime for the starlike dynamics in Hele-Shaw cells. Proc. Am. Math. Soc. **132**(9), 2661–2669 (2004) (electronic)

37. B. Gustafsson, M. Putinar, An exponential transform and regularity of free boundaries in two dimensions. Ann. Scuola Norm. Sup. Pisa Cl. Sci. (4) **26**(3), 507–543 (1998)

38. B. Gustafsson, M. Putinar, *Hyponormal Quantization of Planar Domains*, vol. 2199 of *Lecture Notes in Mathematics* (Springer, Cham, 2017). Exponential transform in dimension two

39. B. Gustafsson, M. Putinar, Line bundles defined by the Schwarz function. Anal. Math. Phys. **8**(2), 171–183 (2018)

40. B. Gustafsson, J. Roos, Partial balayage on Riemannian manifolds. J. Math. Pures Appl. (9) **118**, 82–127 (2018)

41. B. Gustafsson, M. Sakai, Properties of some balayage operators, with applications to quadrature domains and moving boundary problems. Nonlinear Anal. **22**(10), 1221–1245 (1994)

42. B. Gustafsson, H.S. Shapiro, What is a quadrature domain? in *Quadrature Domains and Their Applications*, vol. 156 of *Oper. Theory Adv. Appl.* (Birkhäuser, Basel, 2005), pp. 1–25

43. B. Gustafsson, R. Teoderscu, A. Vasil'ev, *Classical and Stochastic Laplacian Growth.* Advances in Mathematical Fluid Mechanics (Birkhäuser Verlag, Basel, 2014)

44. B. Gustafsson, V. Tkachev, On the Jacobian of the harmonic moment map. Complex Anal. Oper. Theory **3**(2), 399–417 (2009)

45. B. Gustafsson, V. Tkachev, The resultant on compact Riemann surfaces. Commun. Math. Phys. **286**(1), 313–358 (2009)

46. B. Gustafsson, V. Tkachev, On the exponential transform of multi-sheeted algebraic domains. Comput. Methods Funct. Theory **11**(2), 591–615 (2011)

47. B. Gustafsson, A. Vasil'ev, *Conformal and Potential Analysis in Hele-Shaw Cells.* Advances in Mathematical Fluid Mechanics (Birkhäuser Verlag, Basel, 2006)

48. H. Hedenmalm, A factorization theorem for square area-integrable analytic functions. J. Reine Angew. Math. **422**, 45–68 (1991)

49. H. Hedenmalm, B. Korenblum, K. Zhu, *Theory of Bergman Spaces*, vol. 199 of *Graduate Texts in Mathematics* (Springer, New York, 2000)

50. H. Hedenmalm, N. Makarov, Coulomb gas ensembles and Laplacian growth. Proc. Lond. Math. Soc. (3) **106**(4), 859–907 (2013)

51. H. Hedenmalm, Λ. Olofsson, Hele-Shaw flow on weakly hyperbolic surfaces. Indiana Univ. Math. J. **54**(4), 1161–1180 (2005)

52. H. Hedenmalm, Y. Perdomo, Mean value surfaces with prescribed curvature form. J. Math. Pures Appl. (9) **83**(9), 1075–1107 (2004)

53. H. Hedenmalm, S. Shimorin, Hele-Shaw flow on hyperbolic surfaces. J. Math. Pures Appl. (9) **81**(3), 187–222 (2002)

54. H.S. Hele-Shaw, The flow of water. Nature **58**(1489), 33–36 (1898)

55. Y. Hohlov, S.D. Howison, C. Huntingford, J.R. Ockendon, A.A. Lacey, A model for nonsmooth free boundaries in Hele-Shaw flows. Q. J. Mech. Appl. Math. **47**(1), 107–128 (1994)

56. S.D. Howison, Cusp development in Hele-Shaw flow with a free surface. SIAM J. Appl. Math. **46**(1), 20–26 (1986)

57. C. Huntingford, An exact solution to the one-phase zero-surface-tension Hele-Shaw free-boundary problem. Comput. Math. Appl. **29**(10), 45–50 (1995)

58. S. Kharchev, A. Marshakov, A. Mironov, A. Morozov, A. Zabrodin, Towards unified theory of 2d gravity. Nucl. Phys. B **380**(1–2), 181–240 (1992)

59. D. Khavinson, M. Mineev-Weinstein, M. Putinar, Planar elliptic growth. Complex Anal. Oper. Theory **3**(2), 425–451 (2009)

60. D. Kinderlehrer, G. Stampacchia, *An Introduction to Variational Inequalities and Their Applications*, vol. 88 of *Pure and Applied Mathematics*. (Academic Press, [Harcourt Brace Jovanovich, Publishers], New York-London, 1980)

61. I.K. Kostov, I. Krichever, M. Mineev-Weinstein, P.B. Wiegmann, A. Zabrodin, The τ-function for analytic curves, in *Random Matrix Models and Their Applications*, vol. 40 of *Math. Sci. Res. Inst. Publ.* (Cambridge Univ. Press, Cambridge, 2001), pp. 285–299

62. I. Krichever, The dispersionless Lax equations and topological minimal models. Commun. Math. Phys. **143**(2), 415–429 (1992)

63. I. Krichever, The τ-function of the universal Whitham hierarchy, matrix models and topological field theories. Commun. Pure Appl. Math. **47**(4), 437–475 (1994)

64. I. Krichever, A. Marshakov, A. Zabrodin, Integrable structure of the Dirichlet boundary problem in multiply-connected domains. Commun. Math. Phys. **259**(1), 1–44 (2005)

65. I. Krichever, M. Mineev-Weinstein, P. Wiegmann, A. Zabrodin, Laplacian growth and Whitham equations of soliton theory. Phys. D **198**(1–2), 1–28 (2004)

66. O. Kuznetsova, V. Tkachev, Ullemar's formula for the Jacobian of the complex moment mapping. Complex Var. Theory Appl. **49**(1), 55–72 (2004)

67. H. Lamb, *Hydrodynamics*. Cambridge Mathematical Library, 6th edn. (Cambridge University Press, Cambridge, 1993). With a foreword by R. A. Caflisch [Russel E. Caflisch]

68. L. Levine, Y. Peres, Scaling limits for internal aggregation models with multiple sources. J. Anal. Math. **111**, 151–219 (2010)

69. Y.-L. Lin, *Perturbation Theorems for Hele-Shaw Flows and Their Applications* (ProQuest LLC, Ann Arbor, MI, 2009). Thesis (Ph.D.)–Brown University

70. Y.-L. Lin, Large-time rescaling behaviours of Stokes and Hele-Shaw flows driven by injection. Eur. J. Appl. Math. **22**(1), 7–19 (2011)

71. Y.-L. Lin, Perturbation theorems for Hele-Shaw flows and their applications. Ark. Mat. **49**(2), 357–382 (2011)

72. Y.-L. Lin, Classification of blow-up for Hele-Shaw flow solutions driven by suction. Eur. J. Appl. Math. **24**(5), 679–689 (2013)

73. Y.-L. Lin, Classification of degree three polynomial solutions to the Polubarinova-Galin equation. Eur. J. Appl. Math. **26**(6), 849–861 (2015)

74. A. Marshakov, P. Wiegmann, A. Zabrodin, Integrable structure of the Dirichlet boundary problem in two dimensions. Commun. Math. Phys. **227**(1), 131–153 (2002)

75. A.V. Marshakov, Matrix models, complex geometry, and integrable systems. I. Teoret. Mat. Fiz. **147**(2), 163–228 (2006)

76. A.V. Marshakov, Matrix models, complex geometry, and integrable systems. II. Teoret. Mat. Fiz. **147**(3), 399–449 (2006)

77. M. Martin, M. Putinar, *Lectures on Hyponormal Operators*, vol. 39 of *Operator Theory: Advances and Applications* (Birkhäuser Verlag, Basel, 1989)

78. M. Mineev-Weinstein, M. Putinar, R. Teodorescu, Random matrices in 2D, Laplacian growth and operator theory. J. Phys. A **41**(26), 263001, 74 (2008)

79. M. Mineev-Weinstein, A. Zabrodin, Whitham-Toda hierarchy in the Laplacian growth problem. J. Nonlinear Math. Phys. **8**(suppl.), 212–218 (2001). Nonlinear evolution equations and dynamical systems (Kolimbary, 1999)

80. S. Natanzon, A. Zabrodin, Symmetric solutions to dispersionless 2D Toda hierarchy, Hurwitz numbers, and conformal dynamics. Int. Math. Res. Not. IMRN **2015**(8), 2082–2110 (2015)

81. M. Onodera, Asymptotics of Hele-Shaw flows with multiple point sources. Proc. Roy. Soc. Edinburgh Sect. A **140**(6), 1217–1247 (2010)

82. M. Onodera, Stability of the interface of a Hele-Shaw flow with two injection points. SIAM J. Math. Anal. **43**(4), 1810–1834 (2011)

83. A. Petrosyan, H. Shahgholian, N. Uraltseva, *Regularity of Free Boundaries in Obstacle-Type Problems*, vol. 136 of *Graduate Studies in Mathematics* (American Mathematical Society, Providence, RI, 2012)

84. J. Pincus, D. Xia, J. Xia, The analytic model of a hyponormal operator with rank one self-commutator. Integral Equ. Operator Theory **7**(4), 516–535 (1984)

85. P.Ya. Polubarinova-Kochina, On a problem of the motion of the contour of a petroleum shell. Dokl. Akad. Nauk USSR **47**, 254–257 (1945)

86. C. Pommerenke, *Univalent Functions* (Vandenhoeck & Ruprecht, Göttingen, 1975). With a chapter on quadratic differentials by Gerd Jensen, Studia Mathematica/Mathematische Lehrbücher, Band XXV

87. M. Reissig, L. von Wolfersdorf, A simplified proof for a moving boundary problem for Hele-Shaw flows in the plane. Ark. Mat. **31**(1), 101–116 (1993)

88. S. Richardson, Hele-Shaw flows with a free boundary produced by the injection of fluid into a narrow channel. J. Fluid Mech. **56**, 609–618 (1972)

89. J. Roos, Equilibrium measures and partial balayage. Complex Anal. Oper. Theory **9**(1), 65–85 (2015)

90. J. Ross, D.W. Nyström, The Hele-Shaw flow and moduli of holomorphic discs. Compos. Math. **151**(12), 2301–2328 (2015)

91. W. Rudin, *Real and Complex Analysis*, 3rd edn. (McGraw-Hill Book, New York, 1987)

92. P.G. Saffman, G. Taylor, The penetration of a fluid into a porous medium or Hele-Shaw cell containing a more viscous liquid. Proc. Roy. Soc. Lond. A **245**, 312–329 (2 plates) (1958)

93. M. Sakai, On constants in extremal problems of analytic functions. Kōdai Math. Sem. Rep. **21**, 223–225 (1969)

94. M. Sakai, A moment problem on Jordan domains. Proc. Am. Math. Soc. **70**(1), 35–38 (1978)

95. M. Sakai, Analytic functions with finite Dirichlet integrals on Riemann surfaces. Acta Math. **142**(3–4), 199–220 (1979)

96. M. Sakai, The sub-mean-value property of subharmonic functions and its application to the estimation of the Gaussian curvature of the span metric. Hiroshima Math. J. **9**(3), 555–593 (1979)

97. M. Sakai, *Quadrature Domains*, vol. 934 of *Lecture Notes in Mathematics* (Springer, Berlin, 1982)

98. M. Sakai, Applications of variational inequalities to the existence theorem on quadrature domains. Trans. Am. Math. Soc. **276**(1), 267–279 (1983)

99. M. Sakai, Domains having null complex moments. Complex Variables Theory Appl. **7**(4), 313–319 (1987)

100. M. Sakai, Finiteness of the family of simply connected quadrature domains, in *Potential Theory (Prague, 1987)* (Plenum, New York, 1988), pp. 295–305

101. M. Sakai, Regularity of a boundary having a Schwarz function. Acta Math. **166**(3–4), 263–297 (1991)

102. M. Sakai, Regularity of boundaries of quadrature domains in two dimensions. SIAM J. Math. Anal. **24**(2), 341–364 (1993)

103. M. Sakai, Regularity of free boundaries in two dimensions. Ann. Scuola Norm. Sup. Pisa Cl. Sci. (4) **20**(3), 323–339 (1993)

104. M. Sakai, Sharp estimates of the distance from a fixed point to the frontier of a Hele-Shaw flow. Potential Anal. **8**(3), 277–302 (1998)

105. M. Sakai, Small modifications of quadrature domains. Mem. Am. Math. Soc. **206**(969), vi+269 (2010)

106. L. Sario, K. Oikawa, *Capacity Functions*. Die Grundlehren der mathematischen Wissenschaften, Band 149 (Springer, New York, 1969)

107. D.G. Schaeffer, A stability theorem for the obstacle problem. Adv. Math. **17**(1), 34–47 (1975)

108. D.G. Schaeffer, One-sided estimates for the curvature of the free boundary in the obstacle problem. Adv. Math. **24**(1), 78–98 (1977)

109. H.S. Shapiro, *The Schwarz Function and Its Generalization to Higher Dimensions*. University of Arkansas Lecture Notes in the Mathematical Sciences, 9 (Wiley, New York, 1992). A Wiley-Interscience Publication

110. T. Sjödin, On the structure of partial balayage. Nonlinear Anal. **67**(1), 94–102 (2007)

111. B. Skinner, *Logarithmic Potential Theory on Riemann Surfaces* (ProQuest LLC, Ann Arbor, MI, 2015). Thesis (Ph.D.)–California Institute of Technology

112. K. Takasaki, Dispersionless Toda hierarchy and two-dimensional string theory. Commun. Math. Phys. **170**(1), 101–116 (1995)

113. R. Teodorescu, Integrability-preserving regularizations of Laplacian Growth. Math. Model. Nat. Phenom. **15**(Paper No. 9), 14 (2020)

114. R. Teodorescu, E. Bettelheim, O. Agam, A. Zabrodin, P. Wiegmann, Normal random matrix ensemble as a growth problem. Nucl. Phys. B **704**(3), 407–444 (2005)

115. F.R. Tian, A Cauchy integral approach to Hele-Shaw problems with a free boundary: the case of zero surface tension. Arch. Rational Mech. Anal. **135**(2), 175–196 (1996)

116. V.G. Tkachev, Ullemar's formula for the moment map. II. Linear Algebra Appl. **404**, 380–388 (2005)

117. C. Ullemar, Uniqueness theorem for domains satisfying a quadrature identity for analytic functions. Research Bulletin TRITA-MAT-1980-37, Royal Institute of Technology, Department of Mathematics, Stockholm, 1980

118. B.L. van der Waerden, *Moderne Algebra* (Springer, Berlin, 1940)

119. A.N. Varchenko, P.I. Etingof, *Why the Boundary of a Round Drop Becomes a Curve of Order Four*, 3rd edn. AMS University Lecture Series (American Mathematical Society, Providence, Rhode Island, 1992)

120. A. Vasil'ev, From the Hele-Shaw experiment to integrable systems: a historical overview. Complex Anal. Oper. Theory **3**(2), 551–585 (2009)

121. Yu.P. Vinogradov, P.P. Kufarev, On a problem of filtration. Akad. Nauk SSSR. Prikl. Mat. Meh. **12**, 181–198 (1948)

122. F.W. Warner, *Foundations of Differentiable Manifolds and Lie Groups*, vol. 94 of *Graduate Texts in Mathematics* (Springer, New York, Berlin, 1983) Corrected reprint of the 1971 edition.

123. A. Weil, *Scientific Works. Collected papers*, Vol. I (1926–1951) (Springer, New York, 1979)

124. P. Wiegmann, A. Zabrodin, Conformal maps and integrable hierarchies. Commun. Math. Phys. **213**(3), 523–538 (2000)

125. D. Xia, *Analytic Theory of Subnormal Operators* (World Scientific Publishing, Hackensack, NJ, 2015)

126. A. Zabrodin, Matrix models and growth processes: from viscous flows to the quantum Hall effect, in *Applications of Random Matrices in Physics*, vol. 221 of *NATO Sci. Ser. II Math. Phys. Chem.* (Springer, Dordrecht, 2006), pp. 261–318

127. A. Zabrodin, Random matrices and Laplacian growth, in *The Oxford Handbook of Random Matrix Theory* (Oxford Univ. Press, Oxford, 2011), pp. 802–823

128. L. Zalcman, Some inverse problems of potential theory, in *Integral Geometry (Brunswick, Maine, 1984)*, vol. 63 of *Contemp. Math.* (Amer. Math. Soc., Providence, RI, 1987), pp. 337–350

129. D. Zidarov, *Inverse Gravimetric Problem in Geoprospecting and Geodesy*. Number 19 in Developments in Solid Earth Geophysics (Elsevier, 1990)

Index

LECTURE NOTES IN MATHEMATICS

 **Springer**

Editors in Chief: J.-M. Morel, B. Teissier;

Editorial Policy

1. Lecture Notes aim to report new developments in all areas of mathematics and their applications – quickly, informally and at a high level. Mathematical texts analysing new developments in modelling and numerical simulation are welcome.

   Manuscripts should be reasonably self-contained and rounded off. Thus they may, and often will, present not only results of the author but also related work by other people. They may be based on specialised lecture courses. Furthermore, the manuscripts should provide sufficient motivation, examples and applications. This clearly distinguishes Lecture Notes from journal articles or technical reports which normally are very concise. Articles intended for a journal but too long to be accepted by most journals, usually do not have this "lecture notes" character. For similar reasons it is unusual for doctoral theses to be accepted for the Lecture Notes series, though habilitation theses may be appropriate.

2. Besides monographs, multi-author manuscripts resulting from SUMMER SCHOOLS or similar INTENSIVE COURSES are welcome, provided their objective was held to present an active mathematical topic to an audience at the beginning or intermediate graduate level (a list of participants should be provided).

   The resulting manuscript should not be just a collection of course notes, but should require advance planning and coordination among the main lecturers. The subject matter should dictate the structure of the book. This structure should be motivated and explained in a scientific introduction, and the notation, references, index and formulation of results should be, if possible, unified by the editors. Each contribution should have an abstract and an introduction referring to the other contributions. In other words, more preparatory work must go into a multi-authored volume than simply assembling a disparate collection of papers, communicated at the event.

3. Manuscripts should be submitted either online at www.editorialmanager.com/lnm to Springer's mathematics editorial in Heidelberg, or electronically to one of the series editors. Authors should be aware that incomplete or insufficiently close-to-final manuscripts almost always result in longer refereeing times and nevertheless unclear referees' recommendations, making further refereeing of a final draft necessary. The strict minimum amount of material that will be considered should include a detailed outline describing the planned contents of each chapter, a bibliography and several sample chapters. Parallel submission of a manuscript to another publisher while under consideration for LNM is not acceptable and can lead to rejection.

4. In general, **monographs** will be sent out to at least 2 external referees for evaluation.

   A final decision to publish can be made only on the basis of the complete manuscript, however a refereeing process leading to a preliminary decision can be based on a pre-final or incomplete manuscript.

   Volume Editors of **multi-author works** are expected to arrange for the refereeing, to the usual scientific standards, of the individual contributions. If the resulting reports can be forwarded to the LNM Editorial Board, this is very helpful. If no reports are forwarded or if other questions remain unclear in respect of homogeneity etc, the series editors may wish to consult external referees for an overall evaluation of the volume.

5. Manuscripts should in general be submitted in English. Final manuscripts should contain at least 100 pages of mathematical text and should always include

   – a table of contents;
   – an informative introduction, with adequate motivation and perhaps some historical remarks: it should be accessible to a reader not intimately familiar with the topic treated;
   – a subject index: as a rule this is genuinely helpful for the reader.
   – For evaluation purposes, manuscripts should be submitted as pdf files.

6. Careful preparation of the manuscripts will help keep production time short besides ensuring satisfactory appearance of the finished book in print and online. After acceptance of the manuscript authors will be asked to prepare the final LaTeX source files (see LaTeX templates online: https://www.springer.com/gb/authors-editors/book-authors-editors/manuscriptpreparation/5636) plus the corresponding pdf- or zipped ps-file. The LaTeX source files are essential for producing the full-text online version of the book, see http://link.springer.com/bookseries/304 for the existing online volumes of LNM). The technical production of a Lecture Notes volume takes approximately 12 weeks. Additional instructions, if necessary, are available on request from lnm@springer.com.

7. Authors receive a total of 30 free copies of their volume and free access to their book on SpringerLink, but no royalties. They are entitled to a discount of 33.3 % on the price of Springer books purchased for their personal use, if ordering directly from Springer.

8. Commitment to publish is made by a *Publishing Agreement*; contributing authors of multiauthor books are requested to sign a *Consent to Publish form*. Springer-Verlag registers the copyright for each volume. Authors are free to reuse material contained in their LNM volumes in later publications: a brief written (or e-mail) request for formal permission is sufficient.

Addresses:
Professor Jean-Michel Morel, CMLA, École Normale Supérieure de Cachan, France
E-mail: moreljeanmichel@gmail.com

Professor Bernard Teissier, Equipe Géométrie et Dynamique,
Institut de Mathématiques de Jussieu – Paris Rive Gauche, Paris, France
E-mail: bernard.teissier@imj-prg.fr

Springer: Ute McCrory, Mathematics, Heidelberg, Germany,
E-mail: lnm@springer.com

Printed in the United States
by Baker & Taylor Publisher Services